THE PEOPLE OF ASIA SAY NO TO NUCLEAR POWER

THE PEOPLE OF ASIA SAY NO TO NUCLEAR POWER

No Nukes Asia Forum, Japan

Translated from the Japanese by
ann-elise lewallen and Ryan Holmberg

YODA PRESS
79, Gulmohar Enclave
New Delhi 110 049
www.yodapress.co.in

ISBN 978-93-82579-65-6

Editors in charge: Tanya Singh and Nilanjana Bhattacharjee
Typeset by Dharambir Singh
Printed at Replika Press Pvt. Ltd.
Published by Arpita Das for YODA PRESS, New Delhi

Contents

Preface[1]

JAPAN: NUCLEAR POWER EXPORT POWERHOUSE

The first time Japan exported a nuclear reactor body was to Taiwan for its Fourth Nuclear Power Plant, also known as the Lungmen Nuclear Power Plant. The reactor for Unit 1 was manufactured by Hitachi and shipped out from the port of Kure in 2003. Unit 2 was manufactured by Toshiba and departed from Yokohama in 2004. Japan had been quietly exporting peripheral equipment and reactor components for some time. Japan wanted to make money, yet one felt that the country remained cautious about publicly declaring that

[1] The articles in each chapter are compiled from articles originally published in Japanese in the *No Nukes Asia Forum Newsletter*. Interested readers may access individual articles through the catalogue, available here: http://nonukesasiaforum.org.japan/. Special thanks to Kumar Sundaram and Pinar Demircan, for their contributions to chapters 2 (India) and 3 (Turkey), respectively. Thanks also to Japanese language students from the University of California, Santa Barbara, for their contributions in translating the entire text, including Jasmin Augustin, Hiro Bower, Peter Cross, Wona Lee, Dan Luo, Christine Meza, Nelson Moreno, Kaira Nagai, Isabella Pozo, Zhaorong Sun, Yi Lin Tsai, Kirill Viryutin, Minh Vu, Xueyi Wang, Yatong Wang, Lark Yost, Michelle Young, and Elaine Zhang.

it was open for nuclear business. While pursuing customers, it continued to promote its official reputation as 'the only country to suffer wartime nuclear attack'. Since the accident at Tokyo Electric Power Company's (TEPCO) Fukushima Daiichi facility, however, Japan no longer appears to be concerned with maintaining superficial appearances.

The people of the world may find it baffling: 'Why is Japan, the nation that experienced the Fukushima disaster, doing this?' But actually, it is precisely *because* of the Fukushima accident that Japan is pushing forward with nuclear exports. The Japanese nuclear industry is now in its death throes; this outrageous gamble is its only chance of survival.

Prime Minister Abe Shinzō has been promoting his country's nuclear plants throughout the world. 'Having experienced the Fukushima accident,' he says, 'Japan is now able to provide the safest nuclear plants in the world.' Meanwhile, South Korea's president chuckles, 'Japan's declining competitiveness in the global nuclear market offers a great opportunity for South Korea.' Russia assures, 'Our reactor designs are safe, as they are different than those from Japan.' The United States urges, 'In order for the US to start full-scale nuclear exports, Japan must sign the Indo-Japan Nuclear Agreement.' It is as though the Fukushima catastrophe never happened.

We do not mean this figuratively. 'The explosion at the Fukushima power plant was a normal accident, not a radiological accident,' reports the Indian government. It is said that repeating a lie a hundred times makes it true. Indeed, if facts continue to be so blatantly distorted, the people of the world may begin believing in such nonsense. The gauntlet is down. Are we, the No Nukes Asia Forum, Japan, up to redoubling our strength and continuing the fight?

In May 2013, Prime Minister Abe made a tour of the Middle East, followed by a visit to four eastern European countries that June. These were important symbolic trips as they actively promoted the nuclear sales by top Japanese government officials to various foreign countries. New bilateral agreements have emerged as a cause of serious concern; commercial nuclear agreements have been signed with India (signed in 2016), Vietnam, Turkey, and the United Arab Emirates (UAE).

Then Japan's National Diet ratified the Convention on Supplementary Compensation for Nuclear Damage (CSC), which concerns supplementary reparations in the event of nuclear damage. While the ostensible goal of the treaty was to create a network through which signatory nations would collectively cover damages in the event of a nuclear accident, it in fact serves to limit compensation to boorishly small amounts and smooths the way for nuclear exports. Deaf to the pleas of its citizens, the Japanese government has become brazen in its attempts to match the pace of the world's nuclear industry by increasing nuclear exports.

In order to continue building power plants, proponents of nuclear energy are working to create a tight international network with Asian countries like India and Turkey at its center. What can we do in response? We must reach out and get to know people from elsewhere in the world. We must join hands with them and work together to prevent this situation from escalating.

'Nuclear catastrophes have no national borders. Our survival urgently depends on people's solidarity.' This appeal from Kim Won-sik of South Korea bore fruit in 1993 with the founding of No Nukes Asia Forum (NNAF). The first NNAF meeting was held in Japan and welcomed 30 foreign activists from seven countries. Gatherings were held at 28 sites across the country, including in towns neighboring nuclear power stations. With the endorsement of 1,354 private citizens, 177 organizations, and an executive committee of 100 individuals called from across the country commenced the hard labor of organizing NNAF's gatherings. Since then, our organization has held meetings in member nations on a rotating basis almost every year for the past 25 years. Several new networks have appeared alongside NNAF since the Fukushima accident.

Because of the Fukushima disaster and the Japanese government's nuclear export policies that developed in their wake, the global nuclear map is being redrawn with a heretofore unseen speed and urgency. We have to act.

What would happen if any of these reactors were to suffer a severe accident? What has been done so that people can evacuate in the event of an explosion? Are these evacuation plans worth any more than the paper they are written on? They say that it will take another 100 years to fully resolve the crisis at the Fukushima facility. What does the cleanup involve and how much radiation will the workers be exposed to? What hardships have long-term evacuees suffered? What difficulties do parents face in protecting their children from radiation? What have villages and towns done to stop the reckless attempts to restart the nuclear reactors in their backyards? And how can we confront a Japanese government intent on restarting the reactors and advancing nuclear exports?

In the autumn of 2012, three members of NNAF, Japan were denied entry in India at the Chennai airport. They were cross-examined by immigration for two hours. Over and over again, they were asked, 'Do you intend to talk about Fukushima?' 'Do you oppose nuclear plant construction?' Hence we learned the hard way how great and global the impact has been of the words and activism of the Japanese who had experienced the 2011 disaster. We also came to realize that this has in fact been the case for over 70 years, since Hiroshima and Nagasaki.

We must continue to learn about the accident and find our own language to communicate to the world about what happened. We must continue to share daily life experiences of shared sympathy and pain, for through them we better understand ground realities. As people living in a nation that caused a nuclear accident, we should bear in mind our responsibility to share what we know with the world.

We believe that much strength is to be gained by studying the local histories of anti-nuclear movements across Asia and by sharing our knowledge with one another. Just as we in Japan can teach others, what other Asian countries can teach us is practically infinite. For example, the people of the Philippines not only succeeded in toppling the Marcos regime, which had the backing of the United States in pushing injustice to extremes, they also managed to stop the Bataan Nuclear Plant from becoming operational even though its construction was complete. Meanwhile, in Turkey, a country which has no nuclear power plants, anti-nuclear activism has soldiered on for 40 years, fueled by the acute harms caused by radioactive contamination from the Chernobyl disaster and several incidents of exposure resulting from illegally disposed radioactive waste. There is so much strength to be gained by learning from one another.

Since the Fukushima nuclear accident, there have been many studies looking at the actions of nuclear export nations.

Responding to the same situation, the present book sets out to assess the history and present state of anti-nuclear peoples' movements in Asia with a view to opening new prospects for the future. By highlighting how people who have risen up against nuclear weapons and nuclear energy see the world, and by surveying a cross-section of democratic anti-nuclear movements in Asia, we have sought to trace the overall history of the movement and convey the sentiments and character of those who have led the movement. NNAF has been cultivating relations with anti-nuclear movements in Asia since its inception in 1993. We have collected much information, sponsored many joint statements and joint actions, and profiled numerous individuals who are fighting on behalf of their local communities. We hope that this book serves to convey at least some of what we have learned and seen over the many years of struggle.

Chapter 1

Asia's Anti-Nuclear Movements and Japan's Nuclear Exports

Expanding Nuclear Exports in the Post-Fukushima Disaster World

JAPAN'S UNCONTROLLED NUCLEAR EXPORTS

A new system of nuclear exports uniting the world of government and finance has emerged.

Under the mantle of economic growth, a handful of large corporations seeking to turn immense profits are now pushing nuclear exports with rapidity. Under the 'New Growth Strategy' unveiled by the Democratic Party of Japan (DPJ) in 2010, Japan began to pursue nuclear exports. As one prong of this growth strategy, the government sought to craft totalizing 'Export Infrastructure Packages'. Further, nuclear energy was flagged as an ongoing 'critical topic for discussion', at Japan's Conference of Ministries in September 2010. Yet in March 2011, the nuclear accident at TEPCO's Fukushima Daiichi plant drove these plans into a dead end.

Not to be daunted by this situation, the Abe administration which took office in December 2012, raised the banner of nuclear exports under what was termed as the 'Third Arrow' of its economic policy. That is, the prime minister paid visits to several Middle Eastern nations, in pursuit of nuclear technology sales. In his meetings with heads of state, Abe urged them to import Japan's nuclear knowhow. In part, through circulating falsehoods, Abe claimed, 'After the crisis at Fukushima and the nuclear accident, Japan can now provide the world's safest nuclear plants,' or 'Japan has a strong obligation to share the lessons from the accident with the rest of the world'. With utter disregard for whether these target nations had frequent earthquakes, enjoyed freedom of speech, were situated within conflict-prone regions, or possessed nuclear weapons, Japan began to sign bilateral nuclear deals with various nations, one after another.

In early May 2013, Prime Minister Abe made a visit to the Middle East and signed nuclear cooperation agreements with Turkey and the United Arab Emirates (UAE). Mr. Abe also reached a mutual understanding to continue negotiations with Saudi Arabia toward signing an agreement. During these visits, Mr. Abe was accompanied by some 200 representatives from 112 major corporations in Japan. Indian Prime Minister Manmohan Singh visited Japan in late May 2013, and this visit confirmed the reopening of active negotiations toward signing the India-Japan Nuclear Agreement to pave the way for export of Japan's nuclear technology to India. In early June of that year French President Francois Hollande visited Japan and together with Prime Minister Abe announced in a joint statement that the two nations would cooperate in nuclear exports and the nuclear fuel cycle. Then in mid-June Prime Minister

Abe visited the four major Eastern European nations including Poland, and reached an understanding towards cooperation with these nations in the nuclear technology field.

In January of 2014, Prime Minister Abe was invited to India's Republic Day celebrations as the guest of honor. Appealing to the close connections he shared with India's new Prime Minister Narendra Modi, negotiations towards a nuclear cooperation agreement proceeded even further. In April the Lower House formally approved Japan's bilateral nuclear cooperation agreements with Turkey and the UAE. When India's Prime Minister Modi arrived in Japan for a bilateral summit in September 2014, in addition to promoting negotiations for a nuclear power agreement, Mr. Modi signaled his strong interest in bilateral cooperation in the fields of sophisticated military technology and national defense equipment. Based on all of these developments, the Japanese government's move to place nuclear exports as a pillar of its growth strategy for its economic policy going forward is on the verge of running out of control.

The Nuclear Industry: Conspiring to Cross National Borders

Japan has attempted to export nuclear technology since the 1980s. However, in the past the prime minister did not personally engage in overseas boosterism to push the country's nuclear sales, as with the government's present bold assertions that nuclear sales will prop up the economy. Such boosterism is reminiscent of the South Korean government's strategies between 2009 and 2010. Announcing that they would cultivate the nuclear industry towards making it the primary driver of the export economy, following automobiles, ship-building, and semiconductors, the South Korean government proclaimed its goal of exporting 80 nuclear reactors over the next 20 years. During an aggressive trade promotion initiated by President Lee Myung-bak in late 2009, South Korea secured a bid to construct a nuclear energy plant in the UAE.

Japan is currently pursuing joint economic ventures with global industry leaders and attempting to export nuclear technology. In February of 2006, Toshiba acquired BNFL (British Nuclear Fuels Limited) and USA's Westinghouse Electric, placing them under its jurisdiction. From the autumn of 2006, Mitsubishi Heavy Industries commenced joint operations with Areva. In June of 2007, Hitachi and GE (General Electric) merged and established GE Hitachi Nuclear Energy. All of these moves have been directed toward securing their foothold in the world's nuclear market.

The roster of nations with the greatest number of domestic nuclear power plants starts with the U.S., followed by France, Japan, Russia, and South Korea. The demand for nuclear power in each of these nations has already been sated. Hence the nuclear industry hungers for new markets that will ensure its survival. To ensure its survival, the nuclear industry seeks to expand across national borders and around the globe.

The nuclear cooperation agreement required for exports and imports of nuclear material and equipment is defined as, 'an agreement binding two governments, limiting the appropriation of nuclear technologies for military use and prescribing their use thereof for peaceful purposes'. The government then provides its signature and after receiving approval by Japan's National Diet, the agreement goes into effect. The nuclear cooperation agreements that Japan signed prior to the Fukushima Nuclear Disaster include Canada (1960), UK (1968), USA (1968), France (1972), Australia (1972), China (1986), and the EU (2006). Treaties concluded post-disaster indicate the quickening pace in signing bilateral accords and include Kazakhstan (May 2011), South Korea (January 2012), Vietnam (January 2012), Jordan (February 2012), Russia (May 2012), Turkey (June 2014), the UAE (July 2014), and India (2016). Countries currently in mid-negotiation include Thailand, Malaysia, Mongolia, Saudi Arabia, Brazil, Mexico, and South Africa.

The Tragedies that Befall the People of an Export Destination Country

But for Japan, which has yet to resolve the Fukushima Nuclear Disaster, how tenable is it for the nation to attempt to export its nuclear technology to foreign countries? We (NNAF, Japan) doubt much of an explanation is needed. In this text we will instead focus on nations targeted for Japan's nuclear exports and address the specific concerns regarding the planned nuclear export sites, then discuss the strategies that people in civil society have used to resist nuclear imports, and finally consider the specific difficulties in the respective chapters.

First, we will present some of the primary issues of concern. Turkey is an earthquake-prone nation. Up until now thousands have lost their lives to large-scale earthquakes. Sinop, a planned nuclear reactor site, is a picturesque town along the Rias Coast on the shoreline of the Black Sea. Sinop prides itself as one of the most productive fisheries in Turkey, and the people have a strong sense of danger about the possibility of a nuclear disaster; specifically because they worry it would ruin their fishing industry. The Black Sea region has already experienced radioactive contamination from the 1986 Chernobyl nuclear disaster, which reinforces the Turkish people's extraordinarily strong objections towards future nuclear plant construction.

Vietnam, a country currently being targeted by the Japanese government and its corporations for nuclear exports, is ruled by a one-party dictatorship and has no freedom of speech or assembly. Scholars who collect signatures of protest are suppressed. Further, not only are the Vietnamese people unable to raise their voices to oppose government policies; they lack even the means to access accurate information in order to make informed decisions. Japan has secured a bid for constructing its nuclear plant at Ninh Thuan. In the nearby areas many farmers carve a livelihood from harvesting garlic and grapes, and these families are now being forcibly evicted from the land.

India, the country currently being targeted by the world's nuclear industries as the largest nuclear marketplace, is a nuclear weapons state. A situation which can only be described as a tragic showcase of what nuclear imports bring is now unfolding in India.

After developing nuclear weapons by reprocessing the nuclear fuel using reactor technology imported from Canada, India carried out nuclear tests in 1974 and 1998. India has not made any promises of a moratorium on future nuclear tests, and has also only allowed the IAEA (International Atomic Energy Agency) to conduct inspections at certain nuclear facilities. India is now moving full speed towards expanding its nuclear power plants and nuclear weapons production. Herein lies the reason why bilateral nuclear cooperation agreements with India have been maintained to date. As such, nuclear exports from Japanese corporations such as Toshiba, Hitachi, and Mitsubishi Heavy Industries may be complicit in contributing to an ever-expanding nuclear arsenal in India. Nuclear exports themselves are thus directly linked with nuclear proliferation.

In India, there have been several deaths from non-violent citizen protests that were repressed by the police in the communities around the Russian built Koodankulam Nuclear Power Plant and the proposed Jaitapur Nuclear Power Plant which is slated for construction by France. The human rights of the people in an export destination country are ignored and their freedom of speech is stamped out for the sake of importing nuclear technology.

It is becoming increasingly clear that while nuclear exports are protected by the authorities of the partner country, the importing nation bares its fangs towards its own people. Due to the Turkish government's brutal repression of an anti-government demonstration in 2014 which resulted in fatalities, there is concern that similar repression will befall those who oppose nuclear power plants there.

How can we overcome this situation? As a step towards addressing this question, this chapter will broadly outline Asia's anti-nuclear power movement through four periods: the first half of the 1990s, the latter half of the 1990s, the 2000s, and from March 11, 2011 onwards.

An Overview of the History of the Anti-Nuclear Movement in Asia

The Anti-Nuclear Power Movement in its Trajectory Across the Globe: The First Half of the 1990s

In Korea and Taiwan, as democratization movements evolved, questions arose regarding the future of nuclear development and how to handle the nuclear power plants that had been built under dictatorships. The anti-nuclear power movement thus expanded hand-in-hand with democratization. In the Philippines, the struggle to stop the Bataan nuclear power plant—initially constructed under the Marcos regime itself—was inextricably tied to the struggle for democracy. At this time, activities to pitch sales of nuclear technology to Indonesia under the dictatorship of President Suharto increased. These can be said to be straightforward examples of how nuclear power boosters often target countries that lack freedom of expression and the right to protest.

Anti-nuclear movements linked with democratization

During South Korea's military dictatorship, nuclear power plants were constructed in four central locations: Kori, Yeonggwang, Wolseong, and Uljin. After the Great People's Struggles of 1987, the anti-nuclear movement began with these democratization struggles in the background. As a result of the powerful protests organized by residents of these reactor sites, pro-nuclear groups were unable to site new nuclear power plants or nuclear waste disposal facilities. At this time, along with the growing protests to prevent the construction of a third and fourth reactor at Yeonggwang, protests against the construction of nuclear waste disposal facilities spread across the nation.

In the 1991 conflict at Anmyeondo Island, 13,000 out of a population of 16,000 people participated in a protest gathering and demonstration. Resisting an armed police force and eventually driving the authorities into a corner, protestors succeeded in forcing the government to revoke their plans to build nuclear waste disposal sites there. Coined as a battle where 'ants came together in a swarm to fight valiantly and drive away a giant elephant', it was the first page of a new chapter in the history of peoples' movements and served as a precedent for many future actions against schemes to build nuclear waste disposal facilities across South Korea.

In Taiwan, under the 38 years of martial law imposed by the military dictatorship of the Chinese Nationalist Party (Kuomintang (KMT)) wherein protests were forbidden, three nuclear power plants with two reactors each, accounting for a total of six reactors, were built. In 1987, martial law was lifted and the movement against the building of the Fourth Nuclear Power Plant played a pivotal role in the heated democratization battle. The Indigenous Tao community waged an extremely important and steadfast resistance against the transport of nuclear waste to their home island of Lanyu. Separately, the 'Radioactive Apartment' problem arose in Taipei. Victims of radiation poisoning who lived within this apartment building stood up to defend themselves. The radiation originated from contaminated iron building materials sourced from a nuclear power plant. In this incident illegally dumped contaminated iron material was recycled and resold as construction material for houses and roadways, resulting in metal doors and roads that emitted abnormally high levels of radiation. As a result, all of Taiwanese society was shocked at how sloppily this radioactive waste material had been handled. Residents filed a lawsuit, and sought compensation for the harm to their wellbeing.

In Taiwan at the third NNAF meeting in 1995, 30,000 people participated in a demonstration against the planned construction of the Fourth Nuclear Power Plant as well as against France's nuclear tests. With slogans such as 'Down with nuclear weapons' and 'Reject Nuclear Energy' reverberating through the streets, protestors burned models of nuclear weapons and nuclear power plants in the middle of a major thoroughfare in downtown Taipei.

Nuclear power as promoted under Southeast Asia's dictatorships

In Indonesia, plans to establish a nuclear plant in Central Java surfaced in the early 1990s under Suharto's military dictatorship. NEWJEC, a subsidiary of Kansai Electric Power Company, conducted a site feasibility study and concluded that the site was suitable for constructing a nuclear plant. Upon receiving information that Mitsubishi Heavy Industries would proceed with nuclear exports if they secured the bid, NNAF, Japan organized the 'stop nuclear exports campaign' from 1993–1998 and exchanged much information with our Indonesian counterparts. While managing public assemblies, signature campaigns, and various other initiatives inside Japan, we conducted on-site visits in Indonesia with experts on nuclear power and renewable energy. In the midst of a trying situation where the people of Indonesia did not enjoy the freedom to conduct assemblies or demonstrations, they persevered and refused to abandon this movement.

In the Philippines, the fight to permanently shut down the Bataan Nuclear Power Plant, which could very well serve as the symbol of the Marcos dictatorship, continued. During the 'Anti-Marcos and Anti-Nuclear Power' regional general strike of 1985, 50,000 citizens of Bataan filled the center of town and stood firm against the military tanks and armored vehicles for three full days. The movement led to an eruption of people's power in February 1986, which resulted in the downfall of the Marcos dictatorship and led to a freeze on the operations of the completed Bataan Nuclear Power Plant. In 1992, the government announced plans to operate Bataan but later abandoned these plans due to citizen opposition.

OPPOSING THE CIRCULATION OF NUCLEAR EXPORTS: THE LATTER HALF OF THE 1990S

From the mid-1990s the international community began to see Asia's rapidly increasing need for electricity as an opportunity to push for nuclear power adoption, and Japan's nuclear exports strategy came to the fore. Domestic decline in new nuclear power plant orders in Japan also contributed significantly. Commercial nuclear manufacturers felt compelled to shrink their nuclear power divisions, leading to a hazardous situation in which they could not retain staffing levels to ensure maintenance and even minimum engineering capabilities. Such domestic concerns accelerated efforts to promote nuclear exports to nations across Southeast Asia.

Yet, the western-origin nuclear plants that were constructed in South Korea, Taiwan, and the Philippines in the absence of any democratic process, exposed the ugliness of adopting nuclear projects without the people's consent and imposed suffering on the local people. We (NNAF, Japan) had to prevent Japan from making the same mistake at all costs.

The rise of Indonesia's anti-nuclear power movement went hand-in-hand with its democratization movement

The fourth NNAF meeting held in 1996 in Indonesia was staged as an international conference on renewable energy to circumvent political repression from the Indonesian government.

At this meeting attendees came not only from Java, a hub of the anti-nuclear movement, but also from across the archipelago. This meeting spawned the birth of a nation-wide anti-nuclear network. In 1998, through a student-led democratic struggle, the curtain finally came down on Suharto's 30 year dictatorship. The expansion of the anti-nuclear movement in combination with the fiscal crisis of Indonesia's currency at that time brought their nuclear power plans to a grinding halt. The anti-nuclear power movement was thus inseparable from the democratization movement in Indonesia.

Resistance against the revival of the Bataan nuclear project

In the Philippines, President Ramos came to power, and under his (Requisite Infrastructure for Development) plan, he made clear his intention to establish new nuclear plants. In 1996, under the development plan known as 'Philippines 2000', it was revealed that 10 locations had been listed as proposed nuclear plant construction sites. Under this plan, nuclear plants were slated for construction in rural areas, and electricity would be transported to large urban areas through a dizzying network of transmission lines. However, in an island nation like the Philippines, such a pipedream-like plan based on transmitting electricity through an undersea cable exceeding 100 kilometers in length enraged people. Using the connections and leverage cultivated through the Bataan anti-nuclear movement, the group pressured the Philippine Government and put a stop to the nuclear power plant construction plan.

In the fifth NNAF meeting in 1997, after the meeting in Manila, overseas participants arrived by bus to the community neighboring Bataan late in the evening. Despite the fact that it was already past 11:30pm, many local residents awaited the NNAF participants in front of their accommodation, and applauded them as they exited the bus and entered their lodging site. The next day, a crowd of 2000 in Bataan raised blazing torches at a protest against the nuclear power plant. This served to greatly encourage participants inside and outside the country.

The movement against nuclear power plants and research reactors

In 1996, the government of Thailand began to work on a strategy to establish a nuclear plant by coordinating the 'Twenty-one Member Committee' of nuclear specialists. Enlisting the mass media for its nuclear boosterism, the Thai government began to cultivate human resources through newly-established nuclear engineering departments in its universities. They also began organizing study tours, sending relevant officials to visit European nuclear power plants for observation. In 1997, the news surfaced that the Thai government had already conducted feasibility studies and named three locations to host nuclear power stations in Southern Thailand. In response, the anti-nuclear movement broadened its opposition. Moreover, local citizens rose up in opposition to the Ongkharak Nuclear Research Reactor development, which was said to be connected to Hitachi Corporation.

Exporting nuclear reactors from Japan: Taiwan's Fourth Nuclear Power Plant

To the people of Taiwan, the Fourth Nuclear Power Plant was an oppressive symbol that contradicted democracy.

The Democratic Progressive Party (DPP), born from the democratization movement, made an anti-nuclear stance central to its platform. However, with the change in power between the DPP and KMT this stance on nuclear power continuously shifted between promoting and opposing nuclear energy depending upon which party held power. The issue regarding the Fourth Nuclear Power Plant has since become Taiwan's greatest political concern. Despite the unwavering opposition to nuclear power with tens of thousands of people gathering repeatedly to attend enormous demonstrations, Taiwan opened bidding for new nuclear plant construction in 1996. Taiwan awarded the bid to GE, putting Hitachi and Toshiba in charge of manufacturing and exporting nuclear reactors for this project.

In Gongliao District, the site of the Fourth Nuclear Power Plant, the local community continues to wage a fierce battle. In March of 1997, twenty local residents even travelled to Japan to formally discuss their opposition with representatives from the Ministry of Economy, Trade and Industry (METI), Toshiba, and Hitachi. Yang Kuei-Ying, a pivotal member of the opposition concerning the town's plans to construct the nuclear plant, has stated, 'For me it is not sufficient to stop the construction of the nuclear plant here in Gongliao. Rather, I am fighting because I believe that if we can block this nuclear plant here in Gongliao, then we can stop the spread of nuclear power plants throughout Asia'. How did the Taiwanese visitors feel when directly confronting the hypocritically courteous METI representatives and those from Toshiba and Hitachi, who turned them away at the door? As they parted, the METI representative called out, 'Please visit Japan again', to which Yang responded as she smiled, 'I will come again once the plans for the Fourth Nuclear Power Plant are completely shut down'. She has yet to return to Japan.

FIERCE RESISTANCE AGAINST JAPAN'S FIRST NUCLEAR EXPORTS IN THE 2000s

At the start of the 21st century nations across Southeast Asia had hung out the signboards to push new nuclear plant construction, but had made no progress toward bids for or construction of new facilities. During this period, although the catchphrase 'Nuclear Renaissance' was still being used in these campaigns, new plant construction did not actually proceed as planned. Even though nuclear plans continued behind the scenes, the nuclear power industry began to be seen as an industry that had already peaked and was now on the decline. During this time, Japan for the first time exported an entire nuclear reactor body to a foreign nation. Although this was in fact a momentous event, very little was reported in the media.

This was also a time during which India—who did not engage in nuclear commerce with Western nations due to sanctions imposed for its nuclear weapons testing in 1998—was suddenly approached by the USA. After the Three Mile Island accident, USA's nuclear industry could no longer secure new contracts or expand in the United States. This led to them devising plans to export nuclear technology to emerging nations and recognize special provisions for India, a country not signatory to the Nuclear Non-Proliferation Treaty. India quickly became a major battlefield for nuclear exports and began signing nuclear cooperation treaties with countries like USA, France, Russia, and Canada, one after another.

The second invasion: Japan successfully exports a nuclear reactor

Construction of Taiwan's Fourth Nuclear Power Plant began in 1999, but in 2000 the 50-year-long dictatorship of the KMT came to an end, and the DPP's Chen Shui-Bian administration which opposed nuclear plant construction came to power. After three months of public debate on the topic, the Premier (Prime Minister) of Taiwan announced that the construction of the Fourth Nuclear Power Plant would be abandoned. However, just as people were about to celebrate and share the good news, Yang Kuei-Ying, as mentioned earlier, quietly stated that it was still 'Too early to relax'. Three months later, just as Yang had warned, Taiwan's government was thrown into turmoil from an attack by the KMT. At the end of January 2001, the Legislative Yuan voted in favor of continuing construction and in February construction resumed.

Among those joining a sit-in protest in front of the Legislative Yuan, one individual even attempted suicide through self-immolation. The phrase 'Fight with our lives on the line' reverberated through the streets of Gongliao. In 2003 Hitachi's first nuclear reactor was sent out from the port of Kure, and in 2004 Toshiba's second nuclear reactor was sent out from Yokohama. We (NNAF, Japan) worked together with members of the Kure and Yokosuka Peace Squadron and carried out protest actions aboard rubber rafts. Looking up from the small craft we were riding in, the cargo ship carrying the reactor was so massive that we couldn't see it clearly unless we looked straight up. While being tossed about by the large waves we felt resentful of our powerlessness.

The site where the nuclear reactor docked in Taiwan was the same place where the Japanese Imperial Army initially landed to colonize Taiwan. Even today, an anti-Japanese monument commemorating this invasion is erected there. If the nuclear exports are not a second invasion than what else could they be called? Even today, the people of Gongliao still refuse to give up.

AS IF THE FUKUSHIMA NUCLEAR DISASTER HAD NEVER HAPPENED: MARCH 11, 2011 ONWARDS

The Fukushima Nuclear Disaster sent a shockwave across the planet and spurred a renewed movement against nuclear power in every part of the world. As the images of exploding nuclear reactors spread across the globe one after another, countries that had been considering whether to adopt nuclear power announced plans to suspend or postpone new nuclear projects. In countries with nuclear facilities, voices opposed to existing plants and plans for expansion increased in intensity. And in the countries without the nuclear facilities, the public opinion opposing adoption of nuclear power was in the majority. Many wished for the Fukushima Nuclear Disaster to be 'the beginning of the end of the age of nuclear' and those who set their sights on a world without nuclear—either weapons or energy—may have felt that goal to be within their grasp.

However, as if they had been waiting and watching for the opportunity, the pro-nuclear lobby resumed their sales of commercial nuclear technology shortly after the disaster, as if there had never been a nuclear disaster at all. But those who sided with the anti-nuclear position focused on the news of the Fukushima Disaster, aggressively obtaining detailed information and expanding their contact with others in Japan, thus strengthening the opposition movement.

India, the battleground for nuclear exports

In India, swarmed by Japan, Europe, and the US's nuclear energy construction proposals, a hard-to-believe pro-nuclear plan to establish 50 new reactors was set in motion. The nuclear industries of several countries sharpened their claws and refused to pass up this 'business opportunity'. Japan, despite having not yet resolved its own severe nuclear accident, sought to take advantage of the opportunity by merging enterprises such as Toshiba/Westinghouse, Hitachi/GE, and Mitsubishi Heavy Industries/Areva to move forward with a project that would establish an enormous group of nuclear power plants in Kovvada, Mithi Virdi, and Jaitapur. The Mithi Virdi project was eventually denied permission by India's National Green Tribunal and was ultimately relocated to Kovvada in Andhra Pradesh.

Even in Japan, news of the intense battle surrounding the Russian manufactured Koodankulam Nuclear Plant for which construction had been initiated in 2005 and located along the Bay of Bengal in the southernmost tip of India, was accorded prominent coverage. The leader of the Koodankulam nuclear plant opposition movement in Southern India, S.P. Udayakumar, attended the emergency meeting of the 14th NNAF meeting held in Japan in 2011. After visiting Fukushima, Hiroshima, and Iwaishima, he returned to India and conveyed the truth of the 2011 nuclear disaster to the people of Koodankulam, thus spurring the movement even further.

In Koodankulam, several thousand people joined together for a relay hunger strike. In the local community, a total of 56000 people had been detained by the end of 2011. By 2012, armed police increasingly resorted to violence, and frequent raids were reportedly ordered by local authorities but conducted to resemble attacks by thugs. By March, hundreds of protesters were arrested and the presence of armed police had been increased by the tens of thousands. Furthermore, the solitary road that was connected to Koodankulam was blocked and the town's lifelines to water and electricity were cut. Arrest warrants were issued for the leaders of the protest movement who were charged with serious crimes like sedition.

In September, a gathering of 30,000 people filled up the sandy beach and surrounded the nuclear plant. Even though the Koodankulam protest movement was an entirely non-violent protest, armed police forces assaulted the protesters with tear gas and batons resulting in a large number of serious injuries. Despite this, the protesters regrouped once more on the beach, holding hands together in the middle of the ocean and encircling the nuclear power plant, conducting what was termed the 'Non-violent Resistance on Water'. This persistent and powerful non-violent protest received much media attention, thus conveying the local plight not just across India, but to the entire world.

As the movement expanded, a nationwide anti-nuclear movement network came into being. In a country with such an expansive landmass and diverse languages, religious practices, and cultures such as India, the formation of a nation-wide network is something that must be more difficult than we can possibly imagine. Despite this, they managed to succeed, and there is much that we can learn from the Indian activists.

Nuclear power flows toward places which lack democracy

By summarizing history in such a way, it should be possible to put it all together as follows. The nuclear bubble in Japan and Western countries that led to construction of many reactors in these nations popped after the Three Mile Nuclear Disaster (March 1979) in USA and after the Chernobyl Disaster (April 1986) in the Soviet Union (present-day Ukraine), and it became difficult to construct new facilities within these nations.

This resulted in these nations attempting to turn the various countries of Asia into a new marketplace beginning in the 1990s, through schemes to construct nuclear facilities across Southeast Asia in nations such as the Philippines, Thailand, and Indonesia. However, the growth of democracy proceeded in these countries, and the voice of people opposed to nuclear power grew, thus pushing construction in these countries to a standstill. Even in countries under authoritarian dictators, the anti-nuclear power movement grew hand-in-hand with democratization movements, making it difficult for the nuclear industry to continue its expansion.

Following these events, in the 2000s, the pro-nuclear lobbies began turning their sights towards places where freedom of speech and citizen participation in politics had been greatly restricted. USA, France, and Russia, while aiming at the large marketplace in India, also began taking an aggressive approach to developing nuclear markets in Turkey, Vietnam, Jordan, the United Arab Emirates, Sri Lanka, Bangladesh, and Myanmar. After the Fukushima nuclear meltdown, those plans became even more concretized. In the countries that still attempted to import nuclear technology after an accident of that scale occurred, the qualifier of there being an 'energy shortage' has been employed. However anxieties about the proliferation of nuclear arms continue to rise.

Upon closely examining the situation in the countries targeted for nuclear exports, a range of peculiar problems come into view. In each of these countries, the question of how to handle spent nuclear fuel and the lack of freedom of information has also posed serious challenges that require answers. The reason we must call the export of nuclear reactors a crime against humanity is because they directly instigate nuclear proliferation, and are thus equivalent to exporting the worst form of pollution. This also means exporting violence to places whose citizens lack the right to resist.

Asian Anti-Nuclear Movements and Us

LAYING GROUNDWORK FOR NUCLEAR EXPORTS CONTINUES

Under the Abe administration nuclear exports packaged as 'Infrastructure Package Exports' were positioned as the centerpiece of the nation's growth strategy. Up until this point, the Japanese government had come to consistently push nuclear exports across different ministries. Subsequently, METI began to consider the need for a systematic approach to promoting nuclear exports, including everything from human resource training to nuclear plant construction and management. The government has signed onto the CSC which aims to exempt nuclear suppliers from responsibility in case the exported nuclear plants cause an accident. The Prime Minister is personally peddling nuclear sales across the world, resorting to fallacious arguments that make us want to plug our ears, such as 'Due to our experience of the (nuclear) accident, we are now able to provide safe nuclear power'.

The CSC is a treaty which specifies that, in the event of a nuclear accident, if the reparation exceeds the liability limit of the country in which the accident occurred (roughly 47 billion yen), the cost of addressing the accident would be supplemented by funds from a shared pool, donated by each of the signatories to the pact (totaling several billion yen). Japanese Parliament approved this treaty in February 2014, and after Japan's ratification, the treaty was enacted in April. That is, for the CSC to be enacted, it required a minimum of five signatory nations, with a 'combined total energy output from nuclear exceeding 400 GW'. As such, Japan's ratification was essential to the treaty's enactment.

Because the total electricity output of the signatories, before Japan agreed to join USA, Morocco, Romania, and the five states of the UAE, was a little more than three hundred gigawatts, it was reported that USA pressured Japan for early ratification. In comparison to the similar Paris and Vienna Treaties, the standards for various aspects of the treaty are weakened, and the intent of the treaty's design to loosen nuclear export regulations is thus clear.

In the first place, the liability limit is much too low. Based on estimates the damage from the Fukushima nuclear incident may cost upwards of ten trillion yen. But what can be compensated by several billion yen? Further, CSC damage data items are limited to five entries—'death or bodily injury,' 'property destruction or damage,' 'economic loss,' 'the cost of recovery measures,' and 'the cost of preventative measures'—which excludes 'reputation damage' or emotional damage.

Furthermore, nuclear suppliers who manufacture nuclear reactors that cause accidents have become exempt under the principle of concentrating responsibility on operators. The Japanese government has thus focused on including this aspect as meaningful for its participation in the treaty. If Japanese suppliers cause a nuclear accident but are not held responsible, it is not hard to imagine what kind of moral hazards will occur. India, Czech, Lithuania, Australia, the Philippines, Indonesia, Ukraine, etc., have entered their names as CSC signatories. In response to the urgent appeals of activists in India—'Japan, please don't export nuclear plants to India!' and 'Stop the CSC from being ratified!'—we are shocked. Though it hinged entirely on Japan's signing, Japan's ratification has meant that the irresponsible CSC treaty came into effect, and it is already impacting the people of India.

On December 3, 1984 shortly after 12am, a deadly poisonous gas leak occurred in Bhopal, India; tens of thousands of people died and hundreds of thousands of people were injured in this catastrophic accident. Although the accident occurred at a subsidiary factory of USA's Union Carbide company, the management of Union Carbide was not held responsible, nor did they compensate the injured parties. From the experience of this accident, the Supreme Court of India issued a 2010 ruling that 'Foreign corporations that fail to take responsibility for accidents they cause in India will not be tolerated'. Thus Parliament voted to implement the groundbreaking Civil Liability for Nuclear Damages Act. This act includes a clause detailing that, 'When necessary, the Nuclear Power Corporation of India Limited as the operator of India's nuclear plants, will hold the foreign suppliers (providers of reactor cores, etc.) responsible'.

For those countries and corporations that sought to market their nuclear plants in India, this was an exceedingly obstructive clause. Although various limits were pointed out in the Civil Liability for Nuclear Damages Act, the clause enabling the operator to charge foreign suppliers (the corporations providing the nuclear technology) with responsibility was an extremely important part.

However, Prime Minister Singh, who held political power until April 2014, attempted to water down this clause by urging that this phrase be added, 'In order to charge [them] with responsibility, [we] have to prove that there was an intention to induce an accident on the part of suppliers'. In 1998, the Bharatiya Janata Party conducted India's second nuclear weapons test. This political party, led by Prime Minister Modi, took control in 2014 after Prime Minister Singh's Congress Party was defeated in the 2014 elections. Will the Modi administration, with the support of the ruling Bharatiya Janata Party, which espouses a Hindu supremacist agenda in order to create a stronger India and now holds the majority of seats in Congress, now try to sign India onto the CSC? Some observers have suggested that the CSC opens collusion between governments and corporations in export countries in order to advance nuclear exports, and the beginning of a society that irresponsibly sells and spreads nuclear power plants all over the world.

RESTARTS AND NUCLEAR EXPORTS: TWO SIDES OF THE SAME COIN

The struggle to prevent nuclear plants from resuming commercial operation in Japan encourages people of every nation who fight against reactor construction. In May 2012, well-known intellectuals in Vietnam submitted an anti-nuclear petition, including 626 signatures, to the Japanese government. Since Vietnam is pursuing nuclear construction as a national policy, the scholar responsible for this petition drive was threatened by the authorities. This signature-collecting campaign, started during the period when all of Japan's reactors were offline and we learned that this individual was greatly empowered by this situation.

In March 2015, when NNAF, Japan visited the site of Taiwan's Fourth Nuclear Power Plant, local residents rushed to ask us: 'What is the future of the Sendai nuclear plant, which is scheduled to restart? If they allow its restart, the pro-nuclear lobby will become active as well'. The whole world is keeping an eye on the Japanese people, wondering whether the nuclear plants will be restarted and how the Japanese people will attempt to stop the restarts. That is to say, nuclear restarts and nuclear exports are two sides of the same coin. It is no exaggeration to say that stopping nuclear restarts is directly linked to stopping nuclear exports.

There are many examples from the past, when reactor construction projects surfaced, but failed to be implemented and stayed at a standstill. Although these projects may appear to have been stalled by economic crises and political turmoil, we believe that if you look carefully, it is evident that in each case local communities rose up and pushed forward powerful anti-nuclear protest movements, thus hampering these nuclear developments from proceeding further.

After the Three Mile Accident and the Chernobyl Disaster, the nuclear industry slid into complete decline. Today Japanese boosterism is at the heart of the nuclear industry's efforts to recover and resume sales to the rest of the world. At the same time residents of Japan are a driving force for the anti-nuclear movement who have sought to block the actions of the Japanese government. Indeed, the whole world is watching to observe how we in the anti-nuclear movement respond.

WE MUST NOT CEDE OUR FUTURE TO THE NUCLEAR INDUSTRY

It has now been more than eight years since the Fukushima Disaster. With the bilateral nuclear cooperation agreements being cemented in leaps and bounds, Japan's ratification of international treaties, nuclear sales operations conducted by political figures, and nuclear corporations such as Toshiba, Mitsubishi Heavy Industries and Hitachi engaging in meticulous preparation to lay the groundwork for nuclear sales, these shifts might give us a sudden feeling of meaninglessness and a sense of helplessness. It would be easy to be preoccupied by such desperate thoughts knowing about USA's pressure and the enormous power of the global nuclear mafia.

However, we shouldn't allow such thoughts to impede us. Moreover, we should not allow this situation to persist. In the aftermath of the Fukushima Disaster, Japan has become a flag-bearer for promoting nuclear exports, manipulated by the US. Further, if Japan is unable to put a stop on nuclear exports, we will not be able to face the people of the world. Preventing the risks of disastrous accidents worldwide is absolutely essential for human survival.

But what can we do? There is no simple answer to this question. Since we are fighting against a mighty enemy, and the issue involves the entire globe, it is crucial that members of civil society from all fields of knowledge join forces, and continue the fight with determination. People in different parts of the country are committed to oppose the restarting of Japan's nuclear plant, even as they sacrifice their health in devising strategies to achieve this objective.

Unknown facts about the Fukushima Disaster are still emerging and we continue to seek to inform the global community by circulating these. Until now, observers have monitored the financial transactions of Japanese public organizations such as the Japan Bank for International Cooperation (JBIC), Nippon Export and Investment Insurance (NEXI), and Overseas Development Aid (ODA), in financing nuclear exports. Monitoring of public finances earmarked for nuclear exports must be continued. Because Japan exports its nuclear technology to other countries, it is critical that Japanese citizens become involved in examining the following: METI's actions to promote direct support for nuclear adoption in Asian countries; linkages between the Ministry of Foreign Affairs (MOFA) and the International Atomic Energy Agency (IAEA); and the Ministry of Education, Culture, Sports, Science and Technology's (MEXT) allocation of resources to training of Asian nation's personnel to work in nuclear facilities.

There are also plenty of steps to support anti-nuclear efforts that can be taken in everyday life. Opposition to nuclear energy development can be expressed locally as well. Residents can stop buying consumer goods produced by Toshiba, Hitachi and Mitsubishi, and say 'NO' to those corporations that sell nuclear energy. For these corporations who built Japan's nuclear plants, the country is now entering an era of decommissioning. Thus it is clear that ramping up sales of nuclear technology for export does not match the current reality.

For those people in Japan who engage in protest with a view towards a nuclear-free future, and for those across Asia who struggle against nuclear plants, it is crucial that they meet one another, come to know one another, and share the dialogue. That Japanese people fighting against nuclear projects, and those opposed to nuclear in nations with strong pro-nuclear lobbies, meet one another face to face and devise strategies to fight together is crucial. Indeed, these steps are necessary for the collective campaign.

In the myriad and challenging situations confronting the peoples of each of these Asian nations, we are overwhelmed by the creative, thrilling, and intense mobilizations citizens have devised. We would like to urge readers to read each chapter of this book, and familiarize themselves with all the countries involved and the million or more ways Asian peoples have struggled against nuclear energy and nuclear weapons. We would also like the readers and people of Japan to keep seeking the clues and keep on fighting united together as we work toward the goal of making Asia nuclear-free.

Chapter 2

India: Opposition through Non-Violent Direct Action[2]

In the 21st century, India suddenly emerged as a major battleground for the global nuclear power industry's export market. A non-signatory to the NPT (Nuclear Non-Proliferation Treaty), India is a nuclear-armed country that has conducted nuclear tests in 1974 and 1998. The country also possesses all manner of nuclear facilities, from uranium mines to reprocessing plants. Although observers report that serious accidents occur frequently, the Indian government's nuclear operations are shrouded in a veil of secrecy. Information about nuclear technology is not openly shared with the public, on the grounds that the civilian and military uses of nuclear power are intimately linked. Due to the nation's aggressive pursuit of nuclear weapons and nuclear power, people who publicly oppose nuclear projects face an extremely difficult situation indeed.

[2] Special thanks to Kumar Sundaram, editor of DiaNuke.org (Dialogues and Resources on Nuclear, Nature and Society), for his contributions to this chapter.

THE ORIGINS OF THE ANTI-NUCLEAR MOVEMENT

Among the developing nations, India was one of the earliest to embark on the nuclear path. Under Dr. Homi Jehangir Bhabha, a nuclear physicist on close personal terms with Prime Minister Jawaharlal Nehru, India struck out ambitiously on a three-stage nuclear program.[3]

Though voiced only by a small scientific elite, India's nuclear program was beset at the outset by criticism of the plan's scale, the need for massive expenditures, and its secretive nature. Nehru established the Department of Atomic Energy (DAE) directly under the Prime Minister's Office rather than locate it with the other ministries. Scientists like Dr. Meghnad Saha and Dr. Shanti Swaroop Bhatnagar, however, had different ideas. Bhatnagar wanted to develop Calcutta as a hub for nuclear research.

Like most scientists and educated persons of that era, Bhatnagar was optimistic about nuclear energy's potential, and wanted nuclear education to be more open and democratic. He envisioned independent research institutes and universities to spearhead courses and research programs on nuclear technology. Both Saha and Bhatnagar were strongly opposed to nuclear weapons; institutional transparency, they thought, would help ensure against their secret development. Bhabha and Nehru, meanwhile, worked together to leverage their political power in order to create an insulated nuclear establishment.

---

[3] Bhabha devised the three-stage program to utilize India's natural thorium reserves. Under this program, India aims to use uranium-fueled heavy water reactors in stage one, plutonium-fueled fast breeder reactors in stage two, and combine fissile materials with thorium in thermal breeder reactors in the third stage.

In response, Saha sought to find a way to make his position stronger. He stood for election to the national Parliament in 1952, thus enabling him to voice a counter-perspective to Bhabha and Nehru's plans from within the Indian Parliament.

India's nuclear program enjoyed support amongst the general public, as the development of advanced technologies was considered a necessary part of nation building. Nuclear weapons, however, were strongly opposed in the decades immediately following Independence, due to the continuing influence of the non-violent freedom struggle under Mahatma Gandhi. Likewise, the Indian government vocally supported global disarmament at international forums. As a result, it was felt that there was no urgent need for an anti-nuclear movement independent from the government. Most pacifist and Gandhian activists thought the government was on their side.

Thus, when India conducted nuclear tests in 1974 for the first time, it was a shock to most Gandhians, liberals, and leftists. Caught unaware, they had no institutional structure in place to launch a popular campaign against nuclear weapons. Moreover, India had recently fought a war with Pakistan in 1971 and relations between the two countries was still tense. Memories of defeat at the hands of China in 1965 were also fresh. The urgency of national security was felt emotionally by many. Nonetheless, then Prime Minister Indira Gandhi benefited little from nuclear nationalism politically. It failed to raise her sinking popularity. The following year, in 1975, she had to declare a State of Emergency to stave off political contenders.

While a small group of intellectuals criticized the government after the 1974 tests, most date the beginning of India's anti-nuclear power movement to the 1985 protests against the Kakrapar Nuclear Power Plant in Gujarat. The previous year, the city of Bhopal in Madhya Pradesh suffered the world's worst industrial disaster. Late in the evening of December 3rd, 1984, a poisonous gas leak occurred at USA's Union Carbide Corporation. Since the accident occurred in the middle of the night, by the time the evacuation warning was sounded, it was already too late for the local community. The heaviest portions of the toxic gas spread rapidly close to the ground, causing severe injuries to the local poor who were accustomed to sleeping on the ground. Though the death toll for those residing near the factory was estimated to be at least 8,000 people, it is believed that approximately 16,000 to 30,000 died and some 500,000 suffered injuries from the accident. Severe physical problems afflict countless survivors. Victims have not received adequate compensation, nor has the area been fully decontaminated. The heads of Union Carbide returned to the United States without taking any responsibility for the massive disaster they caused in Bhopal.

The Bhopal accident has had an enormous impact both locally and nationally. Industrial risk and corporate responsibility became focal issues for Gandhian activists. Foreign corporations who hoped to operate in India in the future are now required to undergo stringent environmental standards and measures to prevent industrial accidents. Nuclear power plants, the construction of which were additionally seen to be trampling on the rights of the poor, also faced new questions of accountability and safety. It was thus the continuing reverberation of the Bhopal disaster that led to the creation of India's Nuclear Liability Law in 2010.

In 1986, activists launched a massive protest rally against the construction of the Kakrapar Nuclear Power Plant. The government responded with a law banning any assembly involving more than four people. When the police attempted to drive protestors away with batons, horses, and tear gas, thousands of protestors fought back by throwing rocks, escalating the chaos.

As movements against nuclear plant construction began to spread across the country, in 1986 anti-nuclear groups from across India assembled in Mumbai to discuss strategies for collaboration. It was agreed that, while maintaining regular contact with one another and offering mutual support, local groups should be allowed to conduct their campaigns with the tactics that they saw fit.

Properly organized grassroots struggle against nuclear power first appeared in the early 90s in Kerala. A rare, successful battle against the Nuclear Power Corporation of India (NPCIL) was fought in the sleepy town of Peringome, located in Kannur district.[4] The intense struggle is all the more impressive for having been waged at a time when information about the hazards of nuclear technology was not publicly available.

The Communist Party of India (Marxist), which holds power in Kerala, has pursued a strong pro-science and pro-technology agenda. Their modernist-leftist ideology also embraces nuclear energy. The local community in Peringome mobilized them-selves through small meetings, art exhibitions, street plays, and film screenings. Anumukti (Freedom from Nuclear), a

---

[4] Although the Peringome anti-nuclear struggle is more acclaimed, other successful resistance campaigns include those in Bhoothathankettu and in Kothamangalam.

group led by Dr. Surendra Gadekar and Dr. Sanghamitra Gadekar, organized a cycle yatra[5] all the way to Peringome from their base in Vedchhi in Gujarat, a distance of 1280 kilometers. Protests started on 26th April 1990 in Peringome with observation of the anniversary of the 1986 Chernobyl disaster. Agitations continued for months. On August 6th, 1991, in commemoration of the bombing of Hiroshima, government offices across Peringome were picketed by thousands of people. The movement's pinnacle was a massive people's march from Peringome to the district headquarters in Kannur. Many political observers and activists from outside Kerala were struck by the movement's spontaneity and non-hierarchical organization. Although nuclear development had been welcomed by the ruling Communist Party, many key leftwing activists and small leftwing groups supported the opposition. This was a key factor in protecting the grassroots anti-nuclear movement from large-scale police repression.

The success of the anti-nuclear people's resistance in Peringome in 1991 also galvanized protesters against the Kaiga Nuclear Power Plant in Karnataka. Even though plans for Kaiga had been announced several years earlier in the late 80s, successes in neighboring Kerala inspired grassroots struggle against the project. Ultimately the movement did not succeed in stopping the plant from construction. The agitation was led by Dr. Sivarama Karanth under the banner of Citizens Against Nuclear Energy (CANE). Dr. Karanth attempted to make the Kaiga plant an electoral issue when he ran for parliamentary elections.

---

[5] The term 'yatra' embodies the meaning of pilgrimage, and is a campaign strategy which has traditionally been used for protest in India.

Meanwhile, in 1988, India and the former Soviet Union signed agreements to build a power plant at Koodankulam in Tamil Nadu. Strong protests began soon thereafter. In 1989, a massive anti-nuclear demonstration was held in the nearby village of Kanyakumari. Police opened fire, badly injuring three protestors. Due to the protests, NPCIL abandoned its plans to source coolant water from small reservoirs in Pechchiparai and Kodyar. With the collapse of the Soviet Union in 1991, the Koodankulam project was shelved, and remained shelved until 2002. Protests resumed immediately upon its revival, and grew to historic proportions after the Fukushima accident in 2011.

SEVERE ACCIDENTS CONTINUE TO PLAGUE INDIA'S NUCLEAR POWER PLANTS

In 1993, a fire broke out in the turbine room at the Narora Nuclear Power Plant. Workers shut down the reactor as soon as they became aware of the fire. However, they found that the cooling system was not operational. Setting aside concerns about the intense heat and radiation, the workers managed to cool the reactor by transporting water by hand. It was later determined that the fire had been caused by flawed turbine blade design. Companies responsible for the plant's construction had been aware of the problem since soon after ground was broken at Narora, but buried that information to avoid costly changes and delays.

In 1994, the dome covering Unit 1 at Kaiga Nuclear Power Plant (NPP) collapsed, only six months after being installed. The fact that this happened while the plant was still under construction was considered a blessing in disguise. Officials ordered employees to avoid speaking with the press and significantly delayed reporting the accident to the police. The Atomic Energy Regulatory Board (AERB) prepared an accident report based on their investigations. However, since the report was seen as critical of the nuclear industry, elite stakeholders successfully lobbied to have it reclassified as confidential. The Indian government likewise chose against issuing an official statement on the accident.

Public debate on nuclear technologies in India became more open after the Bhopal and Chernobyl disasters. Eminent journalist Praful Bidwai, then Assistant Editor of the *Times of India*, published a series of articles on the functioning of the Department of Energy. In 1986, *Anumukti* (literal meaning: Freedom from nuclear energy), India's first national anti-nuclear magazine, commenced publication. Fueled by the fallout of Chernobyl, it continued to be in print until the early 2000s. Meanwhile, in Bangalore in the early 90s, a group of scientists and activists came together to form an organization called Movement in India for Nuclear Disarmament (MIND).

When the Comprehensive Test Ban Treaty (CTBT) was being prepared for review in the United Nations in 1995, anti-nuclear circles in India discussed the proposal passionately. The Indian government took an absolutist position and denounced the CTBT for not calling for the complete abolishment of nuclear weapons. This resonated with many Indian citizens, who had hoped that nuclear weapons states and their supporters would take concrete action to delegitimize nuclear weapons entirely. Only a small minority in India wished for their country to sign the CTBT. Leftwing groups saw, in the CTBT, an opportunity to reiterate their adherence to nuclear disarmament while denouncing the US hegemony in labelling 'good nukes' and 'bad nukes'.

Meanwhile, the Indian government's financial overseers had turned their attention to the poor performance of their country's nuclear energy program. Amongst the critics was Dr. Manmohan Singh, who as Finance Minister made drastic cuts to the budget of the Department of Atomic Energy. Yet, despite such earlier apprehensions as Prime Minister in 2008, Singh presided over the signing of the India-US Nuclear Cooperation Agreement.

THE IMPACT OF THE 1998 NUCLEAR TESTS

Following the May 1998 nuclear test explosions at Pokhran in the Rajasthan desert, India increased its nuclear development budget by 67%. Until that point, all nuclear power plants in India were constructed for a maximum output of less than 220 MW. Despite the small size of these domestic reactors, difficulty in the initial stages of nuclear plant construction continued to stymie plant operators. India's first test explosions in 1974 led to a long period of international isolation for the country. But as the years passed, advanced nuclear nations in the West gradually warmed to India. Marking as they did the opening of nuclear trade relations with the West and the acceleration of domestic nuclear development, the 1998 tests were a significant turning point for both government nuclear policy and popular anti-nuclear movements in India.

THE TRUE FACE OF THE ANTI-NUCLEAR MOVEMENT: THE CASE OF KAIGA NUCLEAR POWER PLANT

In 1999, participants of the seventh NNAF meeting which was held in India, visited Kaiga. Though protests had been waged against the Kaiga NPP for the preceding 12 years, the local community's perspectives had not been taken into consideration during its construction. Local governments in the Uttara Kannada region of Karnataka passed resolutions to block construction of the Kaiga plant, while India's Supreme Court ruled in 1999 that the siting of the plant should be reevaluated. The DAE and NPCIL, however, ignored both the local resolution and the Supreme Court ruling. Supporters of the plant did not stop there. As part of a sustained program of government repression, 10 fisher folk involved with the opposition were reportedly framed and prosecuted for criminal activities. The Indian government eventually succeeded in splitting the opposition and shutting down further protests from local fisher folk.

Kaiga NPP is comprised of four CANDU (pressurized heavy water) reactors of Canadian manufacture. Since Kaiga is situated inland, to secure a reliable supply of water for the reactors' cooling system, the government constructed a large dam in the scenic Kaiga Valley. From the reservoir's banks, NNAF participants travelled by canoe to the site where three villages had been submerged by the dam's creation. A man from one of these villages told us about how his community has suffered.

> Constructing the dam meant the complete submersion of three villages., Two more villages remain half submerged! The road that we used to use every day has been severed. One confused tiger has begun attacking our water buffaloes. There has also been an increase in mosquitoes and cases of malaria. We are very concerned for our children's future. We would like to move somewhere else as soon as possible, but we don't know if the government would help us with resettlement or finding jobs… we don't know if our lifestyle would be provided for in another place.

Without prior consultation or warning, the government forcibly evicted locals from their lands.

Shrouded in secrecy and executed by force, the case of Kaiga is sadly all too typical of how the Indian government has pursued, and continues to pursue, nuclear development.

JADUGODA: THE URANIUM MINE

Subject to international isolation following the 1974 nuclear tests, India found itself cut off from foreign supplies of uranium. The development of domestic uranium mines became an urgent necessity. They eventually provided the fissionable material for India's nuclear weapons. Domestic uranium being of low quality however, obtaining a sufficient supply required extensive extraction. Not only did local people's health and the environment badly suffer as a result, the safety and rights of those who mined the uranium were entirely ignored by the government.

Between the fifteen villages that lie within a five-kilometer radius of the Jadugoda mines, approximately 30,000 local residents have been exposed to varying degrees of radiation. As there is no fencing around the three tailing dams, children freely play in and around them unaware of the dangers they pose. Villagers traverse this area to gather firewood. Livestock grazes freely on the surrounding grass. Uranium Corporation of India Limited (UCIL), the government-owned company that operates the mines, has taken little care to properly manage radioactive waste. Without treating it, UCIL dumps contaminated wastewater into the Subarnarekha River. Townspeople have historically used water from the river for cooking, laundry, and showering. Collapsed tailings dams, ruptured wastewater pipes, and other accidents have plagued the Jadugoda facility. Much light was shed on the local situation and the hazards of uranium mining by Shriprakash's documentary film, *Buddha Weeps in Jadugoda*, which won the Grand Prize at the Earth Vision International Film Festival in 2000.

In 2015, responding suo moto to media reports, the Ranchi High Court, the state of Jharkand's highest judicial body, directed UCIL to carry out an independent study of its mines' impact on public health. That same year, the USA-based Center for Public Integrity published an important article highlighting UCIL officials' callous and unsafe practices, specifically the disposal of radioactive waste in Subarnarekha River. A major recruitment scan with serious safety ramifications was also unearthed by the Chief Vigilance Officer (CVO) of UCIL in 2017. Ultimately, a further probe by the Central Bureau of Investigation (CBI) was recommended by the CVO.

Starting in December 2000, the Jharkhandhi Organization Against Radiation (JOAR) in Jadugoda has organized nonviolent protests demanding the closure of the uranium mines. Ever since, JOAR has conducted hunger strikes, hosted large gatherings, and stopped trains carrying radioactive waste to prevent additional exposure to radioactivity in local communities. These were the largest protests that Jadugoda had ever seen. Previously, the Jadugoda struggle was considered to be an Adivasi (indigenous peoples) struggle for land rights. Thanks to the 2000 protests, more people learned about the harm caused by the mines. People in the nearby city of Jamshedpur held large protests in solidarity.

When the Meghalaya government was seeking to open uranium mines, *Buddha Weeps in Jadugoda* was translated into the Khasi language to raise awareness among local people. The central government and UCIL had located uranium deposits reportedly 10 times superior to those in Jadugoda at Domiasiat in the West Khasi Hills, located near the Bangladesh border. UCIL petitioned to develop new uranium mines here. However, an alliance had been formed between the peoples of Meghalaya and Jadugoda, and the message of *Buddha Weeps in Jadugoda* had reached a broad audience leading to the formation of a strong protest movement.

In 2003, when UCIL officials visited the planned sites in West Khasi, local residents chased them away with bows, arrows, and swords. Due to the continued threat of violence, the development of mines remained at a standstill. 'There is no reason that the Khasi people should suffer simply because the central government is determined to build nuclear weapons,' stated the Khasi Students' Union, a state-wide activist group. 'The government should put the people who dwell on this land first, and the country's national agenda second.'

As for the newly-planned uranium mine in Jharkhand, the UCIL officials who attempted to visit the site for a public hearing in 2005 were encircled by the local community and physically blocked from entering the venue. The locals shouted in high spirits, 'We are willing to give up our lives, but we will never give up our lands!' Every time a public hearing was scheduled, protesters who learned about the plans swarmed the venue and blocked the hearing from being held.'

As is common across the world, in India uranium deposits and mines are located predominantly in indigenous peoples' lands. Nothing is more important to indigenous peoples than the connection they have with their lands. They will protect singular traditions closely linked to the land, like language, culture, and religious practices, at all costs. UCIL and the government's policy of forcing the indigenous communities to resettle, and abandoning the tailings (i.e. mining slag), waste soil and polluted water remaining from the operation of the uranium mines, deeply impacts the well-being, livelihood, and environment of these communities and is thus inexcusable. Despite such persecution, the indigenous people have continued to fight with determination to stop the development of new uranium mines. Unfortunately, one hears many demoralizing episodes. Some activists, for example, have mysteriously died. But that has not stopped the movement from continuing its struggle.

Concerning the uranium trade: In September 2014, Australian Prime Minister Tony Abbott visited India and concluded the Australian-Indian Civil Nuclear Agreement. Under this agreement, Australia lifted its embargo on exporting uranium to India. During Abbott's visit to India, people from all over India protested, holding up signs reading, 'Go home Abbott!' Protests were also held across Australia, where the government was sharply criticized for agreeing to export uranium to a country with nuclear weapons.

Observers believe that the Indian government plans to use imported uranium to fuel the massive nuclear power stations it intends to import from Japan, France, Russia and the US, while reserving its domestic uranium for nuclear weapons production. Japan's support for the development of nuclear power plants and civilian nuclear technology in India essentially means that Japan supports India's nuclear weapons development.

How the U.S.-India Civil Nuclear Agreement Opened Up a Massive Nuclear Power Market

The U.S.–India Civil Nuclear Agreement triggered a dramatic change in circumstances around the nuclear power industry. USA's interest in courting India had become clear as early as 2004. In July of 2005, then Prime Minister of India Manmohan Singh visited the United States for a summit with then U.S. President George W. Bush, resulting in a nuclear commerce agreement between the two countries. India's international nuclear trade had been hindered by its refusal to participate in the Nuclear Non-Proliferation Treaty or to accept the compensatory provisions of the IAEA. Specifically, the nuclear tests conducted by India in 1974 and 1998 resulted in an embargo on the country's nuclear commerce.

In 2001, the U.S. government abolished the sanctions, and in 2008, the U.S. Senate and the House of Representatives passed the U.S.-India Civil Nuclear Agreement by an overwhelming majority. Under the terms of the deal, the Indian government retains the authority to determine which facilities, materials, or programs are designated for 'commercial use' and which are designated as 'military use' and as such are subject to IAEA inspections.

In September 2008, the Nuclear Suppliers Group (NSG) comprised of 45 member nations, held an unscheduled meeting and approved a 'special exception' lifting the ban on nuclear exports to India. That October, the U.S.-India Civil Nuclear Agreement came into effect. India subsequently signed nuclear commerce agreements with several other countries, including France, Russia, Canada, Britain, and in 2016, Japan. These agreements paved the way for the nation's plans to construct large-scale nuclear power stations across India. Foreign partners in this enterprise include joint ventures linking Westinghouse and Toshiba (in Kovvada), and Areva and Mitsubishi Heavy Industries (in Jaitapur). Rosatom, a Russian company, has been allotted a second site, in Kaveli, Andhra Pradesh, in addition to its previous site in Koodankulam, where four units are slated to join the first two already built. South Korea is reportedly considering investing in nuclear development in India as well.

India's nuclear expansion has also magnified its supportive role in the 'nuclear renaissance' beginning in developing countries. Presently, India is assisting Russia in setting up nuclear plants in Bangladesh and Sri Lanka, and is exploring similar collaborations in other countries of the Global South.

ESCALATION OF THE ANTI-NUCLEAR MOVEMENT IN INDIA

The NSG's 'special exception' triggered alarms among concerned citizens across India. In 2009, citizens convened a meeting to address 'The Political Science of Nuclear Energy and Resistance' in Kanyakumari, on India's southernmost tip. After three days of intense discussion, they established a national network, the National Alliance of Anti-Nuclear Movements (NAAM), and adopted the 'Kanyakumari Declaration' (June 2009). In this declaration, delegates discussed the lack of government transparency on nuclear issues and questioned whether nuclear energy was suited to resolve India's energy crisis. They also addressed health issues that have arisen in areas in close proximity to operational nuclear power plants, and community resistance against nuclear plant construction across regions slated for such development. Finally, the declaration called for an immediate suspension of plans for new reactors, and demanded compensation and medical treatment for those whose health had been impacted by nuclear projects.

On October 2, 2009,[6] NAAM held an anti-uranium mining demonstration in Delhi. Roughly 200 NAAM activists and students lined the streets and performed short plays demonstrating how uranium mining can cause health problems and lead to death. Robust anti-nuclear protests continued in Jaitapur. In November 2009, some 2000 activists and residents from projected nuclear host communities gathered there to demonstrate against nuclear development. Koodankulam participants recounted, 'People across India are rising up against nuclear power. Behind your struggle here, supporters from the entire country stand with you'.

Another participant, from Tarapur, the site of the oldest nuclear power plant in India, railed against the devastating impact of nuclear energy on the marine ecosystem: 'In our area, we had some 350 fishing boats, but now, there's not even one fishing boat left. Fishing has collapsed'. At the close of the rally, participants set fire to a model of a planned nuclear plant, and enthusiastically raised their voices in a collective roar. This initial gathering was the catalyst for the people of Jaitapur and other proposed hosts of nuclear power stations to form a broad-based strike committee. Previously, anti-nuclear protest had remained confined to proposed host communities and the surrounding regions, but in recent years, urban residents have begun to take interest in the anti-nuclear energy movement as well.

[6] October 2nd is Mahatma Gandhi's birthday and is considered one of India's most important national holidays.

In April 2010, the village of Jaspara, in Mithi Virdi, held a rally against nuclear energy packed with 7000 local residents. By June, the police were sent to persuade the local residents to allow government officials to extract soil samples, but the officials were sent back numerous times. Then, on June 11, the company contracted by the Indian government to conduct the study attempted to sneak into the area before daylight with a police escort. However, as the beat of drums reverberated from five separate villages, everything from drawing water, tending livestock, preparing breakfast and the usual morning tasks came to a halt as over 3000 villagers rushed to the proposed site. The police officers unloaded the excavator from the truck, but the people's fierce resistance forced them to reload it and the contractor was compelled to withdraw from the area.

Simultaneously, in the village of Madban, in Jaitapur, more than 1000 people descended on the site of a public hearing. In Jaitapur, some 2300 people have already lost their property due to government seizure of their land, and the majority have not received any form of compensation. The protestors in Madban raised black flags and chanted slogans, filling the public hearing. Over 200 local residents took turns speaking. When an employee of NPCIL claimed that, 'No one was forcibly evicted for the new nuclear power plant', the local residents were enraged. The official hearing was suspended and postponed within an hour, but the angry roars of the people did not subside as they voiced their frustration and chanted slogans for another five hours.

In November 2010, India's Ministry of the Environment greenlighted construction plans for the new nuclear facilities in Jaitapur. When then French President Nicolas Sarkozy visited India in early December of that year, 5000 local residents encircled the planned site at Jaitapur in a human chain, and expressed their strong opposition to the project. Fearing further protests, the authorities implemented evening curfews to restrict the residents' movements. As a result, more than 1500 people were arrested. Going from house to house, the police arrested the leaders of the movement and local celebrities who had expressed solidarity with it. Despite these restrictions, many local residents continued to participate in the protest movement.

On April 18, 2011, during a peaceful protest march at the village of Sakhri Nate, near Jaitapur, police opened fire indiscriminately, and Tabrez Sayekar, a village youth active in the movement, was killed. The killing was met with outrage across the state of Maharashtra, yet no action was taken to hold the police accountable. In response, activists from Maharashtra as well as other parts of India decided to join a yatra, or pilgrimage, from Tarapur to Jaitapur that had already been planned by anti-nuclear activists. The yatra was slated to depart on April 23 from Tarapur, which was the site of India's first nuclear power plant, near Mumbai, culminating in Jaitapur on April 25. However, on the second day of the yatra, most of the activist leaders, including a former navy chief and the former chief justice of Bombay High Court were detained, and the yatra was blocked from proceeding. Participants in the yatra were harassed by the police, and then were detained in the compound of the police station, where they were left in the open air without food for hours. Activists in Jaitapur believe that Tabrez was not the only agitator killed in Jaitapur. Irfan Kazi, another villager from Sakhri Nate, was found dead under suspicious conditions. He had been struck by a police vehicle on his return from bringing his children to school on

December 18, 2010. In another related incident, 72-year-old Pandurang Sadu Barge died of cardiac arrest within 10 minutes of receiving a police notice asking him to furnish Rs. 25,000 as bail for a fabricated charge. Such cases have been brushed aside, in Koodankulam as well as Jaitapur.

Every year since 2012, villagers in Jaitapur have commemorated the anniversary of Tabrez Sayekar's death, and the main road-crossing near Sakhri Nate village has been named Shaheed Tabrez Chowk (Martyr Tabrez Crossing) in memory of the slain activist. Even after acquiring land for the project through unabashed carrot-dangling tactics, the government has not been able to silence the farmers, who insist that they will continue to oppose the project as their land was acquired without their consent.

In June 2010, Japan's *Nikkei* newspaper reported, 'The governments of both France and the United States are urging the Japanese government to sign a civil nuclear deal with India'. Even though companies such as GE in the U.S. and Areva in France had secured contracts to build nuclear power plants in India without Japanese nuclear suppliers, the companies' reactor designs depended on certain components manufactured by Japanese companies. The French and American companies were pushing Japan to wrap up a nuclear agreement with India so that Japanese suppliers could provide the necessary equipment. Japan was being pushed into a conflict between India's nuclear power mega market and public opinion, *Nikkei* reported. In response, many civil society organizations across Japan began working together to oppose passage of the India-Japan Civil Nuclear Agreement (IJNA) by jointly sponsoring petitions, and repeatedly presenting statements and letters of protest to the Japanese government and to India's prime minister. Although opposition to the agreement persists in both nations, in November 2016, the IJNA was signed by Prime Ministers Modi and Abe in Tokyo, and in June 2017, was ratified by Japan's Parliament.

THE IMPACT OF THE FUKUSHIMA NUCLEAR ACCIDENT

The Fukushima nuclear accident shook the entire world. Leveraging the 2008 U.S.-India Civil Nuclear Agreement, the governments of Japan, Europe, and the U.S., together with the nuclear power industry, had heavily invested in India's nuclear market and had been hastening to secure regions in India to develop. In response, India's anti-nuclear power movement had escalated its opposition strategies, and the Fukushima nuclear accident only served to fuel and sustain this resistance. Meanwhile, construction of the two Russian-made reactors on India's southernmost tip, the Koodankulam Nuclear Power Plant, had almost reached completion. In spite of the intense protests there, the government pushed through with construction, and there were reports that the first reactor was nearing criticality.

For this reason, people in the surrounding towns were especially shocked by the Fukushima accident, as it seemed to be a harbinger of a similar danger that might transpire in their own communities. The anti-nuclear power movement in Koodankulam began to develop protest tactics that had seldom been attempted previously. In anticipation of the future utilization of similar resistance strategies by communities targeted for Japan's nuclear exports, such as Jaitapur and Kovvada, we will examine the Koodankulam movement in some detail.

Initial protests against the Koodankulam reactors started when construction plans surfaced in the 1990s, and were ongoing when Russia initiated full-fledged construction on these two reactors in 2005. In 2001, S. P. Udayakumar had organized the group People's Movement Against Nuclear Energy (PMANE), after he concluded his research in the U.S. and returned to his hometown in Tamil Nadu. Under his leadership, the Koodankulam movement has developed original and logical tactics and improved its organizational strategies. In August 2011, Udayakumar took part in the 14th NNAF meeting and visited Fukushima, Hiroshima, and Iwaishima. Upon returning to India, Udayakumar shared what he had seen and heard in Japan with local communities in detail, and in response, the movement in Koodankulam grew explosively.

The community church in the village of Idinthakarai became the center of the protest movement, and on August 16, 2011, some 10,000 residents launched a hunger strike. They also used the church's courtyard as a gathering place, setting up massive tents. Here they held hunger strikes, workshops, and classes teaching about the dangers of nuclear power, and hosted visitors from all over India who met with the local community and held 'relay talks' on an almost daily basis. PMANE also used the Internet to dispatch information to a global audience, including regular updates on the movement they were building in India's southernmost fishing village.

COMMITMENT TO ANTI-VIOLENCE IN THE MOVEMENT AGAINST THE CONSTRUCTION OF THE KOODANKULAM NUCLEAR PLANT

Several thousand people set up a blockade on the main road leading up to the Koodankulam construction site, causing significant delays in the construction schedule. In January 2012, the central government organized a public debate between representatives of both pro-nuclear and anti-nuclear groups. Just prior to the scheduled start of the debate, a Hindu supremacist group had violently attacked local members of the anti-nuclear group, causing several injuries. Since the organizers had been made aware of the planned attack in advance, they had contacted the chief minister's office and had requested and received a firm promise that security would be increased. Despite this promise of security, the attack by the Hindu group was carried out as planned.

Since it was clear that the protestors' safety could not be ensured, the debate was cancelled. Meanwhile, local residents had received word of the attacks, and immediately encircled the nuclear power plant in a massive ring of people. In response to this action by the local community, the central government declared, 'We will not meet with the opposition movement again'. Voices from across India then rang out in criticism of the government: 'How can you attempt to silence a non-violent democratic protest through violence?' By mid-March, the situation had taken a turn for the worse. Although the state government of Tamil Nadu had previously affirmed that it would not force construction of the Koodankulam plant due to residents' strong opposition, this position was abruptly changed and a permit was issued for the startup of operations.

In order to implement the permit, the areas near the proposed plant were surrounded by armed police, and roadways were blocked. The road blockages impeded access to food, electricity and medical treatment for the local communities. According to reports, the armed police stationed in the area numbered in the tens of thousands, and warrants were issued for the movement leaders' arrests. In response to these conditions, Udaykumar and the other protesters began an indefinite hunger strike.

In order to prevent the arrest of the leaders who had been served with warrants, 10,000 protestors formed a protective circle around them. Meanwhile, the police waited at a distance from the protesters. Anticipating that some of the activist leaders would become ill from the hunger strike and would be taken to local hospitals, police intended to intercept and arrest them. As the participants' health began to weaken precipitously, the activists negotiated with the government through a mediator. The state government promised that they would release those villagers who were arrested unfairly and drop the lawsuits against people who had been falsely charged. The indefinite hunger strike was ended by the ninth day, and thereafter was replaced by a relay hunger strike.

The relay hunger strike eventually swelled to include several thousand people. Meanwhile, when the state government failed to keep the promises negotiated at the end of the original hunger strike, activists embarked on another indefinite hunger strike. Moreover, thugs hired by the central government were sent to the village to terrorize the local community, and local residents lived in constant fear. Some 6000 people were charged with crimes such as disturbing the peace, while others were prosecuted under fabricated charges such as attempted murder. At one point, the hired thugs destroyed a school for impoverished children run by Udayakumar. In response to this violence, NNAF, Japan published a joint statement protesting the government's suppression of the anti-nuclear movement, and this appeal was signed by 177 organizations across 22 countries.

Although these protest rallies continued for a year, ranging from several thousand to more than 10,000 people, in September reports reached the opposition that nuclear fuel was being loaded into the nuclear reactors, and activists decided to surround the power plant. On September 9, 2012, more than 30,000 residents gathered along the coastline, encircling the Koodankulam nuclear power plant and raising their voices in opposition. Protestors stayed in position throughout the night, many sleeping on the sandy shore. When the activists awoke on the morning of September 10, a fully armed police force had surrounded them and proceeded to attack the group indiscriminately.

Despite the large numbers of women and children among the protestors, police fired tear gas and beat residents with their batons. Next, the policemen drove the non-violent protestors from the shore and into the sea. In addition, the government blocked off all the roads, severing the residents' lifelines. Furthermore, the government wreaked havoc on the residents' homes, destroying them one after another with a violence that was horrible to witness. They also demolished the assembly hall, which had been erected in the church's courtyard and was used as a base by the protesters.

Despite this miserable turn of events, several thousand people gathered at the shore the next day. Since the police informed the protesters they would be pushed into the sea, they responded that they had intended to conduct their protest in the water from the very beginning. Together they made a human chain in the ocean that they christened the 'Non-Violent Resistance in the Water'. At least three people died during the Koodankulam protests, either directly from police gunfire or due to injuries sustained during the protests. The three men were Sahayam Francis (38-years-old), Antony John (44-years-old) and Ignatius (19-years-old).

Television stations aired a live broadcast of the September 10th attacks, and across India people spoke out and strongly criticized the police violence. Public intellectuals and established writers published protest essays to raise awareness of the indiscriminate violence of the attacks. Many sought to visit Idinthakarai village in solidarity with the protestors. Many overseas activists also expressed solidarity with the people of Koodankulam. However, three Japanese citizens were refused entry to India, as discussed in Chapter 1. Following this wave of protest, India began to reject many entry permits requested by foreigners, especially for those who were suspected to oppose nuclear energy.

The following poem was written as an expression of sympathy with three of the women who were unjustly arrested during the police repression of September 10, 2012, and was translated from the Malayalam language.

Release the Imprisoned Mothers
—V.T. Jayadevan

Sundari, Xavieramma
And Selvi are in jail
In Trichy.
They aren't criminals,
Haven't killed anyone
Haven't stolen anything
Haven't usurped anything
from anyone, haven't insulted anyone,
nor have they ever behaved badly to anyone.
No! Xavieramma, Sundari and Selvi

aren't criminals!
Is it a crime
to say that they have the right
to live without the fear of death;
that they have the duty
to take care of their Mother Sea
who has fed and cared for their tribe
for eons?
Is it a crime to protect the Sea
and the sandy shore
from being burned to ashes?
Is it a crime
to show reverence
which an urban can never have
to a handful of soil
for generations to root themselves
and one pot of water
to quench the thirst of all beings?

Sundari, Xavieramma and Selvi
are still in jail.
Motherbirds whose hearts
forever throbbed
for their children,
for their husbands,
for their dear families;
Women who have
more sense of justice
than any Judiciary;
Women who are
greater than

all great men and women;
Women who have more
historical perspective
than any Epoch-Maker;
Women who have
more wisdom in them
than any ancient wisdom;
Women who have
more honesty in them
than any Truth in the world.

Sundari, Xavieramma and Selvi
are still in jail.
The brutal State
has charged them
with false cases
and imprisoned them.

You have to tell us,
the name of the Book of Justice
based on which
you have accused them
of crimes they didn't commit!
You have to explain to us
the logic of this accusation.
If you and I are citizens of this country,
We are also responsible for this
unlawful crime.
The unjust State
has made all the people
of this country criminals.

Hereafter you shall not
talk big about motherhood,
or the responsibility of mothers,
or the greatness of our ancient culture
or the salt of hard labor
coded in text books.

While these women
who are deeply aware
of the essence of life,
of eternal justice,
of the ultimate responsibility
of every living being on Earth,
who have lived a life
in tune with these values
are still in jail.

While these women
languish in the jail,
while they weep
like motherbirds
caught in the nest of a hunter,
thinking about their burning forests,
their life-giving sea,
and crying for their children,
do we have the right
to be indifferent and
to be sleeping
in our comfort zones
like this?
(Translated from the Malayalam by S. Santhi)

Despite these creative expressions of opposition, Unit 1 of the Koodankulam nuclear plant nevertheless came online in June 2013, with Unit 2 following in February 2016. However, the Koodankulam community's historic opposition will doubtless be carried onward to influence protest movements in Jaitapur and Kovvada. Needless to say, the people in Koodankulam have certainly not abandoned their movement, and continue to express their opposition to this day through various forms of protest. One of their initiatives, known as 'Train Yatra', offers an unusual example of protest action within India's anti-nuclear movement. In 2014, a group of people boarded a train from Tamil Nadu, which is located in southern India, and tied an anti-nuclear banner to the side of the train. At each stop, they disembarked, delivered speeches and distributed materials, continuing in this fashion all the way to the northernmost regions of India. As mentioned earlier, the term yatra embodies the meaning pilgrimage, and is a method of campaigning which has traditionally been employed in India.

Despite commissioning of the Koodankulam nuclear facilities with much fanfare and brutal repression of local protests, the two units have barely functioned since attaining criticality in 2013 and 2016, respectively. The plant has seen an unprecedented number of shutdowns and 'trips', which trigger a shutdown. Activists in Koodankulam and independent experts have alleged since 2011 that substandard equipment was used in constructing the reactors, owing to a deception perpetrated by the Russian supply company. Their suspicions were confirmed by the 2017 report of India's Comptroller and Auditor General (CAG), which has indicted NPCIL for financial losses and delays, as well as the purchase of defective turbine rotors.[7]

In 1999, at the 7th NNAF meeting, we listened to a report about a similar style of yatra, the 'Cycle Yatra'. In this form of yatra, 20 to 30 people rode bicycles to local communities to inform the residents of the dangers of nuclear weapons and nuclear power through public assemblies and distribution of flyers. Occasionally these cycle yatras spanned distances as far as 1000 kilometers. Even as we have mastered the use of social-network sites as a way to organize protest in Japan, there is much to learn from the steadfast approach of India's anti-nuclear movements, lessons which will help us to improve upon our own strategies.

India is now in the main stage in the battle over nuclear power. How will Japan respond to the current situation there? Will we overlook the police repression and military pressure, even as deaths and casualties among the protesters increase? Will Japan force the construction of Japanese-manufactured nuclear plants across India?

We propose that we must link hands with India's anti-nuclear energy and anti-nuclear weapons citizen groups. We need to cultivate a better understanding of what is occurring at both planned sites and at current nuclear power plants in India, and we would like to share what's happening in Japan with our colleagues in India. It is necessary for us to take a closer look at the outcome of the debates over India's Nuclear Liability

[7] Report of the Comptroller and Auditor General of India on the Koodankulam Nuclear Power Project (KKNPP), Units I & II (38 of 2017), Presented to Parliament, 27 December 2017.

Law. We need to create movements that will definitively block the implementation of the India-Japan Nuclear Cooperation Agreement.[8] Even though our collective enemies may appear to be giants blocking our way, the knowledge we have gathered can help us to implement strategies and plans of action that will lead us to overcome these challenges.

[8] Translator's Note: The Indo-Japan Nuclear Cooperation Agreement was signed by Prime Ministers Modi and Abe in November 2016, and ratified by Japan's Parliament in June 2017.

Chapter 3

Turkey: Jolted by the Memory of the Chernobyl Accident[9]

RADIATION FROM CHERNOBYL REACHES TURKEY

The 1986 Chernobyl nuclear disaster carried severe pollution to all of Europe. Radioactive contamination of food and the environment became serious problems. In Japan, this unprecedented calamity led to an upsurge in the anti-nuclear power movement and increased concern about the safety of imported foodstuffs. Under the Law on Food Hygiene, Japanese regulators set provisional radioactivity limits at 370 becquerels per kilogram of imported food. Many European imports were found to exceed these limits, and were returned to their countries of origin. Looking at the list of rejected items, alongside imports from France and other European countries, one finds things like nuts and leaves from Turkey. At that time, most Japanese consumers only had a

[9] Special thanks to Pinar Demircan, writer and editor of yesilgazete.org (Green News) and co-ordinator of nukleersiz.org, for her contributions to this chapter.

vague notion that foodstuffs from Turkey might also be a cause for concern.

Despite the 2011 Fukushima Daiichi nuclear accident, in 2013 Prime Minister Abe Shinzō visited Turkey in a bid to export Japan's nuclear technology to Turkey. After signing the first of Turkey's intergovernmental agreements with Russia in 2010 and its second in 2013, Erdoğan began to promote nuclear power as indispensable to Turkey's energy needs. Shortly after the deal with Russia, Turkey and Japan entered into Japan's first intergovernmental nuclear cooperation agreement since the Fukushima disaster. When Japan's plans to construct a nuclear power plant in beautiful Sinop, on the shore of the Black Sea, became public, anti-nuclear groups in Turkey and Japan began to work more closely together. It was through this networking that we of NNAF, Japan, came to realize how deeply Turkey had been harmed by Chernobyl, and how the pain and anger had driven the people of Turkey to rise up against the building of nuclear plants in their country.

40 Years of Opposition to the Akkuyu Nuclear Plant

Way back in 1955, Turkey signed a bilateral nuclear cooperation agreement with the United States. Turkey thus sought to establish a nuclear energy program from very early on. Various obstacles, however, have so far prevented even a single nuclear power plant from being built in the country. In 1956, the government created the Turkish Atomic Energy Authority (TAEK), which joined the International Atomic Energy Agency (IAEA) in 1957. The first research programs related to nuclear power generation commenced activity in 1965. The

Turkey Electric Production & Supply (TEAS) company was founded in 1970, and initiated feasibility studies in 1971.

Located on the coast of the Mediterranean Sea, Akkuyu emerged as the leading candidate for Turkey's first nuclear plant in 1976. Over the past four decades, there have been four occasions on which it seemed like the bidding process was going to result in the construction of the facility, but in each case the project has ultimately failed. While political difficulties have also been a factor, the strength of opposition movements has been the main reason the Akkuyu plant is yet to be built.

With Chernobyl, Turkey's anti-nuclear movement became unshakable. The radioactive plume from the destroyed Chernobyl plant passed directly over Turkey. Five days after the accident, the air radiation dose in the Western part of the Black Sea was 20 times higher than normal. In Turkey's western region of Thrace, it was 1000 times higher than normal. While contamination was widely reported in various Eastern European countries, the impact on Turkey remained unknown due to the Turkish government's failure to authorize an official investigation.

Large numbers of Turks first experienced the horrors and consequences of nuclear accidents with a spike in cancer diagnoses on the coast of the Black Sea following the Chernobyl disaster. Then we also learned the fact that the Turkish government had allowed about 50,000 tons of contaminated tea leaves to be mixed with 130,000 tons from the previous year's harvest, which were then released into the market. Hence, it is not surprising that the same people who succeeded in obtaining information on the harms of the Chernobyl accident should also be so opposed to the construction of nuclear power plants at home.

CANCELLATION OF THE THIRD BID

In October 1997, Turkey opened an international bidding competition for new nuclear power plant construction. The companies that cast bids did so through one of three consortiums: 1) the NPI Consortium, which included Siemens (Germany), Framatom, (France), GecAlsthom (France), Campenon Bernard (France), GarantiKoza İnşaat (Turkey), and Tekfen Construction (Turkey); 2) the AECL Consortium, which included Atomic Energy of Canada, Hitachi (Japan), Itochu (Japan), Gama Industry (Turkey), Guris Construction (Turkey), Bayindir Construction (Turkey), Ansoldo (Italy), and Daiwoo (South Korea); and 3) the Westinghouse-Mitsubishi Consortium, which included Westinghouse (USA), Raytheon-Duke Energy (USA), Mitsubishi Heavy Industries. (Japan), Enka Construction (Turkey), and MNG (Turkey). The winner of the bid was to be announced in June 1998.

Statements and studies were promptly issued by Canadian anti-nuclear organizations. They argued that Canada should not export nuclear technology to Turkey for a variety of reasons. For one, it was unethical for Canada, which had not allowed the commissioning of any new nuclear plants domestically since 1978, to try to reap profits by exporting nuclear technology to Turkey. Also, Canada was taking on significant economic risk by agreeing to loan Turkey the large sum it required to construct a nuclear power plant. Corruption inside Canada's federal nuclear energy corporation, Atomic Energy of Canada was also cited. As Pakistan's development of nuclear weapons in the early 80s showed, Turkey had contributed directly to global proliferation in the past . The domestic situation with the Kurds and security tensions vis-à-vis Greece over Cyprus, they also argued, pointed to a history of human rights violations and domestic political instability.

In May 1998, Turkey's Minister of Energy announced that the winner of the bid would be delayed. Around the same time, an influential retired military official stated in a local television interview that Turkey needed to develop its nuclear potential in order to counter nuclear threats from nations such as Israel, India, Pakistan, and Iran. His comments generated controversy both domestically and abroad, for they suggested that Turkey was pursuing nuclear energy as cover for developing nuclear weapons.

Meanwhile, public protests heated up in Akkuyu. One concern was how the plant would impact the region's biodiversity, and specifically the lives of the endangered monk seals which inhabit Akkuyu Bay. The high frequency of earthquakes in Turkey was another issue that turned people against nuclear power. A series of massive jolts had just struck the country in 1997 and 1998.

Between December 1998 and January 1999, Istanbul suffered a series of radiation exposure incidents caused by the illegal dumping of medical cobalt-60. A lead canister containing the substance had been transported from Ankara to Istanbul, where it languished in a warehouse for nine months. Then, the warehouse shifted hands. Unaware of its radioactive contents, the new owners sold the canister as scrap metal, where its crushing initiated a chain of contamination incidents that ultimately harmed some 300 people. Seven children and eleven adults had to be rushed to the hospital after suddenly developing symptoms from radiation exposure. One person died; another had to have their fingers amputated. The incident sent shock waves across Turkey.

In August 2000, the Turkish prime minister announced the cancellation of the Akkuyu Nuclear Power Plant bidding process. Behind this decision was a magnitude 7.4 earthquake that had struck the city of Izmir, located 500 km from Akkuyu, on August 17 of the previous year. 17,000 people died, while half a million lost their homes. Prompted by strong anti-nuclear sentiment amongst the populace, both members of the national Parliament and leaders of the city of Mersin, the closest municipality to Akkuyu, expressed their strong opposition to the plans for power plant's construction.

REACTIONS TO THE FUKUSHIMA NUCLEAR DISASTER

In May 2010, plans for the Akkuyu facility were revived when Turkey and Russia signed a bilateral commercial nuclear agreement. The local anti-nuclear movement in Akkuyu, which had previously been described as engaging in 'concentrated fire' upon its enemy, was there to receive the challenge. Then, after the Fukushima Daiichi accident in 2011, what had largely been a local movement suddenly exploded nationwide. In Mersin, which is near Akkuyu, protestors formed a long human chain stretching 159 km and spanning 30 towns between Akkuyu and Mersin, and leading to the cancellation of public hearings that were to be held in Mersin.

Turkey experienced another contamination scandal in 2012. Toxic materials, including radioactive substances, were discovered in a deserted lead plant in Izmir, the country's third most populous city. Suspicious of increased cases of cancer and other ailments, area residents contacted reporters. In 2014, Turkey's Green Party opened a court case against the TAEK in response to the illegal dumping of radioactive waste on the plant's premises.

As the plant was located near a residential neighborhood, children passed by it almost every day on their way to school. Though radiation levels are high, authorities have still done nothing more than erect a fence around the plant. A number of other cases of illegally dumped radioactive waste poisoning the environment have been reported from across Turkey. While this has yet to be proven, some suspect that what is being dumped is illegally imported radioactive waste.

JAPAN EXPORTS THE SINOP NUCLEAR PLANT

Seemingly oblivious to such events, Prime Minister Abe visited Turkey and signed the Turkey-Japan Nuclear Cooperation Agreement in 2013. This was the first of Japan's intergovernmental agreements to participate in nuclear power plant construction since the Fukushima Daiichi disaster. As critics noted, the biggest problem with the agreement was the insertion of a clause stipulating that, provided that both governments agreed in written form, Turkey would be permitted to engage in uranium enrichment and nuclear fuel reprocessing within its own national jurisdiction. It defies the imagination as to why Japan would want to provide one of the largest military powers in the Middle East with the ability to extract and process plutonium.

The reactor planned for Sinop is the Atmea 1, a new model of nuclear reactor. The reactor is being marketed and built by the Atmea Joint Venture consortium, which is led by Mitsubishi Heavy Industries and the French Nuclear Power Corporation Areva. Members of the Sinop Anti-Nuclear Platform offered the following in response: 'How sad that Japan is trying to build a nuclear power plant after suffering so much from the Fukushima Daiichi accident'. As this criticism comes from direct victims of Chernobyl, Japan should be truly ashamed.

Sinop is a beautiful city on the coast of the Black Sea. The city is blessed with a rich variety of flora and fauna, and boasts a vibrant tourist industry. The vast majority of Turkey's fisheries are based in the Black Sea, with Sinop reportedly pulling in the biggest catches. For the purposes of building the Akkuyu plant, the Turkish government has already acquired 60 square kilometers in this region. In a mere 10 square-kilometer area, and they have already clear-cut some 650,000 trees, even though the land has not been leased yet.

In its attempt to build a nuclear power plant, Turkey has joined hands with the two nations responsible for the world's largest nuclear accidents, Russia and Japan. The irony of such a partnership is sobering. More and more people from across Turkey are wondering, 'How can Japan export nuclear technology while its own people seek to stop the operation of nuclear power plants at home? How totally unethical.' Active since plans for the Akkuyu nuclear plant first surfaced, Turkey's anti-nuclear movement grew substantially after the Chernobyl disaster, leading to, the formation of the Turkish Anti-Nuclear Platform in 1993, which hosts frequent debates, meetings, and other events in opposition to nuclear power.

In 2006, in response to the Turkish Atomic Energy Authority's announcement that Sinop was the planned location for Turkey's next nuclear plant, and in commemoration of the 20th anniversary of the Chernobyl accident, 15,000 people protested in Sinop. This demonstration is said to be the largest mobilization around an environmental issue in Turkey's history. On April 26, 2014, marking the 28th anniversary of the Chernobyl disaster and the 3rd anniversary of the Fukushima nuclear accident, 10,000 people from all over the country congregated in Sinop to participate in another large protest against nuclear power. Similar meetings were held in Mersin, close to the Akkuyu site, and even in the capital Ankara.

On April 1, 2015, the Turkish Parliament approved a Japanese contract for the Sinop nuclear power plant. Ground-breaking on the facility's four reactors, which are to have a combined generating capacity of 4480MW, is slated for 2017[10]. Meanwhile, the cornerstone-laying ceremony at the Akkuyu site was held on April 14, 2015. Akkuyu Nuclear Power Plant has been the subject of ongoing legal challenges. After a court ruling in which the government supported the company, challenges to the plant's legality are now being heard in the Constitutional Courts with rulings anticipated in 2019. Total costs for the four reactors at the Akkuyu facility, with a combined generating capacity of 4800MW, are projected at about 20 billion dollars. They are slated for completion by 2020. On April 25 of the same year, 30,000 people at a protest in Sinop cried, 'No nuclear power' and 'Hey Japan, stop nuclear exports!'

A Country Called Turkey

According to a public opinion poll, 64% of Turkey's population is opposed to nuclear power. In the villages where nuclear plants have been proposed, opposition exceeds 80%. Even

[10] On December 4, 2018, Mitsubishi Heavy Industries, which led the consortium, decided to abandon the Sinop Nuclear Project due to soaring costs.

many people who have no direct contact with the anti-nuclear movement are against the development of nuclear power. Lawyer's associations, doctor's associations, and other groups and individuals with social influence have made their opposition to nuclear power clear.

In August 2013, immediately after Prime Minister Abe and President Erdoğan concluded their nuclear agreement with huge smiles, a public demonstration that had originally focused on deforestation of the trees and shrubs of Istanbul's Gezi Park was met with unbelievably violent police suppression. Over 3,000 people were arrested, and thousands were injured. 12 people lost their eyesight and 11 people were killed.

Erdoğan came to power as the prime minister in 2003, before becoming president in 2014. Ever since then, Turkey has witnessed a daily decrease in tolerance while the general social atmosphere becomes more oppressive. In July 2016, a so-called military coup transpired, further emboldening Erdoğan to suspend human rights and control daily life in Turkey. As freedom of speech is increasingly curtailed, it is not rare to find cases of publishers being pressured to dismiss dissenting reports and even of journalists being imprisoned. According to the NGO, the Committee to Protect Journalists, for two years running, in 2012 and 2013, Turkey had imprisoned more journalists than any other country in the world. Meanwhile, Turkey has also become militarily involved in the Syrian War; some four million refugees from Syria have flooded into Turkey. Thousands of academics who oppose the war have also been marked by the government and some have been dismissed from their university posts. Since the anti-nuclear power movement is bound to increase in intensity in the years to come, many fear that they will be subject to violence like that inflicted on protests at Gezi Park in 2013.

In 2017, Erdoğan proposed a referendum to increase his presidential authority and simultaneously lessen the power of Turkey's legislative branch, the Turkish Parliament. In April 2017, the referendum resulted in Erdoğan gaining expanded authority, with 51% of votes in favor of the referendum. Voting observers criticized these results, citing that 1–2% of the pro-referendum votes should have been deemed invalid due to being submitted in unstamped envelopes. Despite public outcry, appeals to the High Election Council to look into the possibility of voting fraud, and recommendations from the European Union commission to conduct an investigation, the High Election Council accepted the questionable ballots. Erdoğan was thus able to amend the Constitution and increase his authority as president.

Turkey is home to a complex, multiethnic, multireligious, and multilingual society, the product of a long history of demographic shifts that transcend national boundaries. We, in Japan, have a hard time grasping the complexity of Turkey's history and its relationships with its neighbors. We cannot forget the words of a certain Turkish scholar in response to our inquiry on how Turkey's anti-nuclear power movement has maintained momentum for such a long span of time.

Countries that are multicultural, multireligious, and multilingual are the norm across the world. Japan's homogeneity is the exception. Isn't it precisely because people feel in their hearts that nation-states and national boundaries are constantly changing that their first priority is to protect the place where they live no matter what?

Chapter 4

Vietnam: Promoting Nuclear Power Under a One-Party Dictatorship

STRENGTHENING NUCLEAR TIES BETWEEN JAPAN AND VIETNAM

> (In Vietnam) there are scientists but there are no engineers—establishment of legislation is yet to come and the knowledge of international frameworks is inadequate. However, there are advantages in provincial Vietnam. The strong political system of one-party rule, coupled with the nation's enthusiasm for hard work, labor regulations, and high standards for knowledge, contributes to rapid policy implementation and distribution of resources, once policy has been decided.
> (Tetsuya Endo, August 1999)

This is an excerpt from a report that Tetsuya Endo, then vice president of the 13th Southeast Asian Mission of the Japan Atomic Industrial Forum (JAIF) to Vietnam, contributed to the *Genshiryoku Sangyou Shimbun* (Nuclear Industry Newspaper) in August 1999. Although the article received little attention upon its initial publication, even today it is a shock that the parties planning nuclear exports to Vietnam would express their intentions so brazenly. In the same article, Mr. Endo stated that if Vietnam set up a preparation committee to consider introducing nuclear energy and requested his cooperation, he would assist them gladly, writing that 'JAIF could be deployed as a point of entree to create an all Japan system to integrate private and public sector involvement' as one proposal. From this point on, Japan and Vietnam's cooperative effort to introduce the development of nuclear power in Vietnam progressed rapidly.

In December 1999, JAIF and Vietnam's Atomic Energy Commission agreed to a 'Memorandum of Understanding to cooperate in preparation for implementing nuclear power generation in Vietnam'. In March of 2000, both parties signed 'The Basic Plan for the Year 2000'. Under this plan, both parties confirmed that they would cooperate with the activities of the five subcommittees that had been established under the Governmental Steering Committee for Nuclear Power, housed within the National Atomic Energy Council. These five subcommittees included personnel training, site selection, preparation and adjustment of legal codes, research and development, and public relations. At the same time, the Japan Vietnam Nuclear Cooperation Liaison Committee brought together representatives from the electric industry, plant manufacturers, trading companies, and research institutions, and the framework for expediting nuclear plant construction in Vietnam was established.

On their end, JAIF took such actions such as accepting trainees from Vietnam and inviting key stakeholders and government personnel to Japan in order to create a wider understanding of nuclear power. In 2006, METI hired a group of public-private joint representatives and established support for the training of nuclear technology personnel. In 2008, METI and Vietnam's Ministry of Commerce and Industry signed the 'Nuclear Cooperation Document', confirming support for preparation, planning and promotion, training of personnel, safety regulations, and public relations campaigns for nuclear plant construction. By 2009, they had established the 'International Nuclear Cooperation Council', confirming cooperation with the project from both the public and private sectors. The Federation of Electric Power Companies of Japan designated Japan Atomic Power Company as the body responsible for conducting the feasibility study for the Vietnam nuclear project. Beginning in 2002, the Japan Electric Power Information Center created a training program and began—and continues to this day—to accept trainees from Vietnam to fill nuclear power plant operation and management positions. Indeed, nuclear power in Vietnam is All Japan.

Japan Accepts the Order for Vietnam's Second Nuclear Power Plant

Vietnam's plans to introduce nuclear power progressed rapidly, and various countries such as Russia, China, Korea, France, and Japan began actively promoting nuclear commerce to Vietnam. In 2010, Prime Minister Kan Naoto and Prime Minister Nguyễn Tấn Dũng convened a bilateral summit in Hanoi. Dũng announced subsequently that Vietnam and Japan would partner to construct Vietnam's second nuclear plant. The site proposed for the reactors was the village of Thai Anh, in the Ninh Hai district of central southern Vietnam's Ninh Thuan province. In February 2011, it was announced that Japan Atomic Power Company and Vietnam Electricity Holding Company (EVN) had signed a cooperation agreement towards the introduction of nuclear power.

The following month, in March 2011, the Fukushima nuclear accident happened. On March 29th, Deputy Prime Minister Nguyen Thien Nhan gave the following response: 'We will learn from the Fukushima accident, this will allow us to construct even safer nuclear plants together with Japan and Russia'. Precisely the opposite response was heard from nuclear planning authorities in countries such as Thailand, Indonesia, and China, who decided, albeit temporarily, to postpone any plans for implementation or promotion of nuclear technology.

In September 2011, it was revealed that the International Nuclear Energy Development of Japan Co., Ltd, a company founded for the export of Japan's nuclear power plants to Vietnam, had signed a Memorandum of Cooperation with EVN. That October, Prime Minister Noda and Prime Minister Dũng held bilateral talks at the Prime Minister's Residence in Tokyo, reconfirming the 2010 bilateral agreement that Japan would move forward with construction of nuclear plants in Vietnam. Following the talks, the Japan Atomic Power Company initiated its feasibility study in Vietnam.

Vietnam had proposed the following six conditions for nuclear imports from Japan: 1) Japan would provide the most technologically advanced and safest reactors already in operation in Japan; 2) Japan would support Vietnam in developing its nuclear industry; 3) this support would extend to personnel training; 4) Japan would allocate funding support; 5) Japan would supply fuel in a safe manner; and 6) Japan would provide support for a system of nuclear waste management.

The question remained as to exactly what kind of support Japan would be providing to manage the nuclear waste produced by Vietnam's plants. In May 2011, plans were announced for the Comprehensive Fuel Service (CFS), which would export uranium fuel, as well as accept nuclear waste produced by other nations' nuclear reactors. Many began to conjecture that Japan had proposed the condition of accepting the spent nuclear fuel and exporting it to Mongolia, raising voices of concern that Japan had made the proposition in order to win the heated bidding wars for Vietnam's second nuclear plant.

The Japan-Vietnam Nuclear Cooperation Agreement was ratified by parliament and went into effect in December 2011. Some confusion resulted from 12 abstention votes cast by legislators from the ruling Democratic Party of Japan (DPJ) during the ratification process. By this time, many civil society groups in Japan were in opposition to this hasty nuclear agreement and had set up campaigns to block it. Their statements pointed out some of the problems with exporting Japan's nuclear reactors to Vietnam:

> METI has budgeted two billion yen for the feasibility study and it has been carried out, but the study's findings remain undisclosed. The Ninh Thuan nuclear power plant is sited next to a national park, which is protected as a hatching ground for sea turtles. There are also concerns surrounding both the construction and operation of the nuclear power plant. In a prior incident, the Japanese government had provided funding through its Official Development Assistance for the Can Tho Bridge, which collapsed while under construction, causing multiple deaths. There have also been several accidents in which hydroelectric dams released water without any warning to those downstream, leading to many deaths and

other injuries. Corruption is another serious issue and many have criticized Vietnam's governance as ineffective. Tsunami measures are also unclear, and three seismic faults have been pointed out in the nearby region. The provincial capital of Phan Rang is within 20 kilometers from the proposed nuclear plant location. Phan Rang's population is 180,000 and there is no clear evacuation plan. Vietnam's one-party rule dictatorship also restricts and censors information and freedom of expression. Opposition to Vietnam's nuclear plans arising from neighboring countries such as Thailand also cannot be ignored.

RISKING THEIR LIVES TO OPPOSE NUCLEAR ENERGY

In May 2012, a petition was submitted to the Japanese government, signed by a group of 626 members of Vietnam's civil society such as Nguyen Xuan Dien.[11] The petition spelled out the group's opposition to Japan's nuclear exports to Vietnam, and argued that Japan, despite its advanced technology, would be unable to prevent a nuclear accident. The authors further pointed to the fact that many in Japan have strong reservations about nuclear power as well. After noting that all of Japan's commercial nuclear power plants had been placed in shutdown mode, they admonished the government of Japan, stating, '[Given all of the above], for the Japanese government to fund the construction of nuclear power plants in Vietnam is irresponsible, inhumane, and immoral'. Shortly after this petition was sent, a group claiming to represent injured soldiers barged into Dien's research instititute and threatened him, demanding that he remove the petition.

[11] Nguyen Xuan Dien is a scholar of the classical music of Vietnam and is a researcher based in Hanoi.

Further retaliation followed, as Dien was subpoenaed to report to the Hanoi authorities; following that his public blog was shut down.

Under one-party rule, it was an anomaly for anyone to sign a petition openly. Those who signed the petition did so with the knowledge that it might endanger them and would, at the very least, disadvantage them. It would not be an overstatement to say that they had risked their lives in signing this petition. Their open signatures came as a shock for many. Dien explained, '[People] were encouraged by the ongoing shutdown of nuclear power plants in Japan'.

A contributor to the NNAF newsletter under the penname 'N', wrote:

> In Vietnam, under the one-party rule of the communist party, all assemblies, association, expression, and reporting are strictly limited. Even after being summoned by the authorities, Dien's punishment was limited to a fine. After this, he revived the blog and reported this information there, including the police summons and the fine. In response to the news, donations from people empathizing with his situation flowed in. Ultimately, the donations exceeded the amount of the original fine. In actuality, unauthorized blogs such as Dien's are only accessible to urban intellectuals and those who have regular use of computers and the Internet. Only authorized media such as newspapers, television, and radio, reach most of the general public. Vietnamese intellectuals opposed to nuclear power thus pleaded, "Inside Vietnam, we cannot mobilize freely, and we do not have the ability to stop nuclear power plant construction. We call upon the Japanese citizens to use your strength and please stop the export of nuclear reactors to Vietnam".

This book's main objective is to communicate the sentiments and desires of the citizens of various countries who have acted to oppose nuclear technology. However, for this chapter on Vietnam, it was difficult to provide exact quotes or first-hand commentary. In Vietnam, there are no public spaces in which acts of protest are allowed to take place. Even if they feel compelled to act, people who would raise their voices in protest are unable to since doing so would jeopardize their safety. Furthermore, due to government's censorship, the public is denied access to accurate information and knowledge. For the Japanese government to exploit Vietnam's one-party rule in order to export nuclear reactors is irresponsible and unethical in the extreme.

STRATEGIES FOR HALTING JAPAN'S NUCLEAR EXPORTS TO VIETNAM

However, Japanese scholars of Vietnam studies are bravely raising their voices against Japan's nuclear exports to Vietnam. Through astute analysis of the academic literature on Vietnam's politics, history, and society, many scholars have exposed the problems associated with nuclear exports to Vietnam. These scholars concur that it is difficult even to approach regular citizens of Vietnam in order to obtain their opinion when conducting studies. Despite the difficulties in communicating, they urge that it is critical that the flow of information continue to reach Vietnam's engaged citizenry regarding developments from the Fukushima accident and Japan's rejection of nuclear energy.

Focusing on the village of Thai Anh, the projected site for the nuclear plants to be constructed by Japan, a Japanese film-maker produced the documentary *Stealth Nukes* (*Shinobiyoru Genpatsu*). The film reflects the filmmaker's gentle personality, and one scene in particular is unforgettable: A female farmer harvesting grapes proudly boasts that '[in this village] crops are abundant'. She is unaccustomed to being filmed, so when the camera is directed towards her, she laughs shyly along with her female co-workers, finally laughing so hard that she falls over backwards. The women burst with laughter in the midst of their work, their voices echoing together across the placid scenery of the beautiful farm village.

These communities apparently have been at the receiving end of propaganda claiming that '[Japan's] Fukushima nuclear power plant has been restored. The accident is a thing of the past. Japan is going to use the Fukushima nuclear accident as a learning experience to construct the most advanced and the safest, latest generation nuclear reactors'. The pro-nuclear lobby's reiteration of these lies is sinful and unacceptable.

Even within the Vietnamese government, the pro-nuclear lobby is not a monolithic power. In a 2009 National Assembly resolution, the number of parliamentarians who voted against nuclear power reached double digits. This is an anomaly in Vietnam. Furthermore, in 2012, Guen Guam, Vietnam's Minister of Science and Technology, officially announced that 'Due to the lack of personnel, it is necessary to postpone construction (of the nuclear plant)'.

On January 2014, Prime Minister Dũng suggested that the ground breaking for the first nuclear power plant might be delayed from 2014 to 2020. Concerning Russia's bid for the first Ninh Thuan nuclear power plant, Dũng remarked that 'It is essential to guarantee maximum safety as well as the most economically efficient nuclear energy generation, but if we cannot achieve both, we will abandon construction'.

In September of that year, it was officially decided that construction would be delayed for a period of 7 to 9 years.

However, in June 2015, Prime Minister Dũng approved the relocation and resettlement plan for the nearby residents of two sites located in Ninh Thuan province. Relocation for the first plant is estimated to impact 2,084 individuals, and for the second plant, 4,911 individuals. Under the current plan, resettlement is slated for completion in 2018.

Dien was emboldened in his petition activism by the fact that there were no operational nuclear power plants in Japan at that time, and it is clear that the nuclear situation in Japan creates considerable influence over nuclear exports to Vietnam. If we continue to move forward with anti-nuclear activist movements within Japan, our efforts will eventually reach the people of Vietnam.

† † †

What if the unthinkable occurred...

—Q. Mai Ly

Then at that moment, what would the wind do?
It would probably continue to innocently blow, as if nothing occurred at all.
What would the Jin River do?
It would probably lose its way, but nonetheless, continue to flow.
Then at that moment, what would farmers do?
They would probably frown, look up to the sky, and stop their hands from cultivation.

What would the wealthy ones do?
They would probably collect their fortunes, and quickly escape.

What would the poets do?
They would probably compose a poem, sorrowfully, and be enraged.
What would researchers do?
They would probably bury themselves in studying a sculpture that had just been excavated.
What would professors do?
They would probably continue teaching the same exact course as last year with all their heart.

Then at that moment, the sea... What would the sea do?
It would probably tremble furiously with rage for a brief second,
but then, it would stop immediately
And then it would continue to bellow in an unchanging melody for eternity

The above poem was written by Q. Mai Ly, a female poet who lives in an indigenous Cham village located 17 kilometers from the planned nuclear site in Japan. After Ly signed Dien's protest petition, security authorities came to her house to interrogate her. Nevertheless, her poem managed to pass the censorship review, and the anthology of poetry in which it is included was published in 2015.

Chapter 5

Indonesia: The Joint March to Democratization and the Anti-Nuclear Power Movement

PLANS FOR CONSTRUCTION OF THE MURIA NUCLEAR POWER PLANT

Interactions between Indonesia and Japan's anti-nuclear movements started in the early 1990s. The evolution of the Indonesian government's plans for the Muria Nuclear Power Plant over the course of the 90s is a microcosm of the problems surrounding nuclear exports. The Indonesian government first proposed a nuclear power park on the Muria Peninsula, on the northern side of central Java, in 1989. With seven to 10 reactors, the park was to have a generating output of seven gigawatts by 2015. President Suharto, who was in the midst of a 32 year dictatorship, is said to have ordered the project.

An international bid for a feasibility study took place in August of 1991. NEWJEC Inc., a consulting firm that is a subsidiary of The Kansai Electric Power Company, beat rivals from Europe and North America for the contract. The proposed site was in the village of Barong, located on the western side of the Muria Peninsula. At the time, Barong had a population of 7000. Surrounded by large, state-run coconut and gum plantations, the village subsisted half on agriculture and half on fishing. As the village had never been blessed with electricity, people lived solely by the light of kerosene lamps.

NEWJEC's feasibility study was carried out in Barong between 1991 and 1996. Environmental groups attempted to hold public conversations about the problems associated with nuclear energy, only to be repeatedly foiled by the refusal of local authorities to issue permits. Protests staged by activists in front of the Jepara Regional Assembly were broken up by the military. In order to prevent activists from interacting with local residents, outside visitors were prohibited from staying overnight within the region surrounding the proposed nuclear site. All visitors were required to report to the security forces. Barong village was encircled with a barbed-wire fence, forcing residents to enter and exit through a checkpoint manned with armed soldiers. Despite such conditions, environmental organizations from across Java succeeded in disseminating information about nuclear power's dangers. Their influence was such that intellectuals and religious leaders also began voicing opposition to the government's plans.

Immediately after being awarded the feasibility study, NEWJEC applied to Japan's Ministry of Finance and the Ministry of International Trade and Industry for permission for a service agreement under Japan's Foreign Exchange Control Law as well as for trading insurance. Both applications were successful. The company also applied for a loan from the former Japan Export & Import Bank (JEXIM), which is now the Japan Bank for International Cooperation (JBIC). JEXIM decided to make the loan based on 'Individual Judgments on Feasibility Loan and Construction Loan'. Out of the 1500 million yen total cost of the feasibility study, JEXIM paid nearly 900 million. The 1993 Feasibility Report concluded, 'Of all the sites in Indonesia, Muria Peninsula is the best suited for the construction of a nuclear power plant'. The final, 1996 Feasibility Report came to the same conclusion. Criticism from environmental organizations was severe: 'The report ignores all negative analyses. It's pure formality, nothing more.'

THE 'STOP NUCLEAR EXPORTS CAMPAIGN'

Together with multiple citizen groups, in 1993 NNAF, Japan helped initiate the Stop Nuclear Exports Campaign. Amongst the many actions we organized were citizen assemblies and signature-collecting campaigns. On numerous occasions we accompanied Japanese experts, on subjects like renewable energy, on trips to Indonesia to tell people there about the dangers of nuclear power amongst other issues. Also, every year, we invited anti-nuclear activists from Indonesia to Japan, so that we could stay updated about local conditions in Indonesia and deepen contact between our groups.

Meanwhile, concerns began to mount over whether JEXIM would finance the construction of the nuclear power plant itself, the contract for which was likely going to be awarded to Mitsubishi Heavy Industries and Westinghouse, as it had the feasibility study. The Stop Nuclear Power Exports Campaign declared, 'It is unacceptable to use public funding to finance nuclear-related business given unresolved questions about the industry's safety. Stop financing now'. Petitions were submitted to JEXIM and the Ministry of Foreign Affairs while signature-collecting campaigns continued. Even Japan's Diet conducted three investigations. Ultimately more than 150,000 people signed their name to a petition demanding, 'Don't Use Public Funding to Export Nuclear Power to Indonesia'. This petition, along with the names of 80 members of parliament, was submitted to the chairmen of both houses of the Diet, asking the Japanese government to deny use of public funds for nuclear exports. In December of 1996, the Foreign Affairs Committee of the Upper House vowed, 'The Foreign Ministry will not use Official Development Assistance (ODA) funds for nuclear exports'.

HOLDING THE NO NUKES ASIA FORUM, JAPAN

It is often said that nuclear power and democracy are incompatible. Like elsewhere in Asia, nuclear power in Indonesia is a symbol of military dictatorship and economic-industrial autocracy. Anti-nuclear organizing was carried out in the face of extremely difficult circumstances and doubled as a pro-democracy movement.

The fourth NNAF meeting was held in Indonesia in 1996. Under repressive conditions in which meetings and demonstrations cannot be held freely, mounting a conference is itself an act of great significance. What's more, on July 27th, the day before the forum was to commence, the military secretly organized an attack on the headquarters of the Democratic Party of Indonesia. More than 100 people were killed, and a so-called 'riot' broke out. The atmosphere was extremely tense. The forum nonetheless dared to open in Jakarta, Solo, and Yogyakarta. People from Java, the traditional center of the country's anti-nuclear movement, of course attended. But so did supporters from islands like Sumatra, Lombok, and Sulawesi, marking the birth of a nationwide anti-nuclear network in Indonesia.

Unfortunately, in February 1997 Indonesia's lower house approved revisions to the national nuclear energy law that had been in existence since 1964. Under the revised law, formal approval was no longer required from Indonesia's legislature on matters relating to nuclear energy. Consultation alone was now deemed sufficient. The only parliamentarian to oppose the bill was a female member, Mire Puriyongo, who had previously visited Japan on invitation from the Stop Nuclear Exports Campaign. Her house was subsequently searched by the authorities.

Despite the new law, public opposition to nuclear power continued to spread, leading to plant construction delays. The following year, in 1998, Suharto's military dictatorship was brought to a close by a student-led people's struggle. The onset of Asia's currency crisis meant further setbacks to Indonesia's plans for nuclear power.

SEVERAL PROPOSED SITES

In 2000, the price of petroleum skyrocketed worldwide. In Indonesia too, the government raised alarms about the dangers of relying on oil for generating electricity. Memos were exchanged with South Korea and Russia on the subject of collaborating on the construction of nuclear power plants. A Comprehensive Energy Investigation Committee was established, reopening debates about introducing nuclear energy into the country.

This was all suspect from the very beginning. Why would Indonesia, as one of only a few oil- and natural gas-producing countries in the world, want to go out of its way to import uranium to generate electricity? Nevertheless, one government official after another had something to say about nuclear power.

One such development was a plan to construct a nuclear power plant on Madura Island, located off the eastern end of Java. With a feasibility study already completed, it was reported in 2001 that Korean Electric Power Company had been working with Indonesia's National Nuclear Energy Agency on plans to build a nuclear plant for the purposes of desalinating seawater and generating electricity. Local environmental groups and plant site residents joined forces and declared, 'We have no desire for Madura Island to become home to a nuclear tragedy'.

The Indonesian government's long-term energy plan of 2006 clearly stated that nuclear power facilities capable of generating 4GW would be completed by 2024. The Minister of Energy said that the country's first nuclear reactor would be up and running in Java by 2015. Like Japan, Indonesia is located on the volcanic, Pacific Ring of Fire and is thus prone to seismic activity. Knowing how little their government had done in response to natural disasters and pollution problems in the past, the Indonesian people only felt more threatened by these new developments.

Around the same time, plans for Russian-made, floating nuclear power stations off the coast of the Sulawesi Islands in Gorontalo province were made public. Of the many shady activities that came to light was the story of thirteen top Indonesian officials visiting South Korea and Russia for tours of nuclear power facilities.

MURIA NUCLEAR POWER PLANT IS HARAM

In 2007, due to the resurgence of government promotion of nuclear power, the opposition movement in Muria Peninsula took off. That June, thousands of residents, artists, and celebrities demonstrated in Jepara. The following week 3000 people protested in Kudus, which is next to Jepara. In July, Nurudin Amin, head of the Java branch of Nahdlatul Ulama (NU), a national Islamic organization, visited Japan and South Korea, two likely candidates for nuclear exports to Indonesia. Amin and his supporters submitted petitions to the Japanese government and Japanese corporations, held public meetings in Tokyo and Osaka, and went to Hamaoka (site of a nuclear power facility) to network with local residents. In South Korea, they worked with citizen groups and demonstrated against Korean Electric Power Corporation.

On the night of August 31st, 6000 men, women, children, and senior citizens from Barong village set out on a long march. Their destination was the headquarters of NU's Central Java division, 35 kilometers away. On the following day, 100 NU clerics were to open discussion on what position the organization was to take on the matter of nuclear development. The villagers wanted to ensure that their opposition was heard.

At the end of a long debate led by Nurudin Amin, NU approved a fatwa (an ordinance based on Islamic religious law) declaring that the planned plant at Muria was 'haram', that it was prohibited under Islam. 'The latent dangers of nuclear power are greater than the benefits it brings. After discussing the matter thoroughly, we have concluded that the Muria Nuclear Power Station is against Islamic law. Radioactive waste is far too great a risk for a mere 2–4% of our energy needs.' Considering that, of the Jepara regency's total population of approximately 1.2 million people, one million are NU supporters, it is easy to imagine how great the social impact must have been when the organization took a strong stance on this issue.

The anti-nuclear movement in Barong Village was not led by activists from the cities. Rather, local villagers linked up with university students from central Java and shaped the movement on their own. Students came to the conclusion that there were more important things to do than attend classes. Typically, at any one time, 10 students lived in the village to support local residents' activities, taking turns to show their face on campus.

Persistent Pro-nuclear Parties

Despite these developments, in November 2007 the Japanese and Indonesian governments signed the 'Cooperation Document on Aid for Nuclear Energy Development, Preparation, Planning, and Promotion'. Meanwhile, anti-nuclear organizers were frequently assailed with threatening emails. Victims reported the intimidation to Indonesia's National Human Rights Committee. Testimonies including phone calls threatening, 'Be careful, or we'll kidnap and shoot you'. There were also cases of pro-nuclear parties showering localities with money, then threatening people if they refused to give their support. Japan should be ashamed for advancing policies that essentially encouraged such actions.

In February 2008, thousands of villagers held a protest march in Barong, before literally barricading the Nuclear Energy Agency's local office and weather tower. After village elders and members of parliament piled stones, people brought cement, sand, stone, and bricks to the site in their cars, creating a wall four meters high and blocking entrance to the facilities.

That August, Indonesia's Minister of Research and Technology unexpectedly announced, 'A nuclear power station is going to be built in Banten province in western Java'. The Nuclear Energy Agency added, 'Preliminary studies are being carried out at two sites in Banten', while 'plans for development on the Muria peninsula continue'.

Risky Bangka Island

Not long after, yet another new candidate site suddenly surfaced: Bangka Island, which floats off the southern coast of Sumatra. A feasibility study was carried out as part of the Five-Year Plan announced in 2010. Bangka Island is the same size as Okinawa Island, with a population of about one million. While agriculture accounts for the largest portion of the island's economy, it is one of the few places in the world with reserves of tin. Byproducts of tin refining contain uranium, thorium, and other radioactive materials. Poor management of slag at government-run tin refineries had been an issue for years. The likelihood of health hazards and environmental pollution were very high. But according to reports, since workers feared being fired, there was little chance of cases being brought against the industry.

It was on such an island that two new nuclear sites were being proposed. A number of Japanese had visited Bangka over the past few years. They reported that locals were told by Indonesia's Nuclear Energy Agency that 'the Fukushima nuclear incident has already been resolved. There is absolutely no problem'. Also: 'Japan's nuclear power plants are based on old designs. The new models to be built here are absolutely safe'. Locals nonetheless expressed their anxieties: 'Our ancestors have lived here for many generations. We have no desire to move and live elsewhere'; 'The Nuclear Energy Agency only repeats how safe nuclear power is. Since we don't trust them, we studied the matter ourselves'; 'We harvest rice and vegetables from the fields. We are not rich, but nor are we poor. We don't want nuclear power plants'; 'Most fishermen are against nuclear power'. Although a large-scale oppositional movement has yet to materialize on the island, solidarity with the Bangka peoples is essential considering how the nuclear energy industry runs rampant on the global stage these days.

As for People Who Live in Nuclear Export Countries

In 2006, Indonesia and South Korea announced a 'Declaration of Strategic Partnership', strengthening the possibility of South Korean nuclear exports to Indonesia. The Indonesian people responded as follows: 'When plans of nuclear exports from Japan emerged in the 1990s, Japanese came to Indonesia to explain the problems of nuclear energy. This greatly helped local residents form educated opinions on the matter. If today there is a greater possibility of South Korean exports, then it is imperative we act together with the Korean people'. Solidarity between the citizens of different nuclear export countries will undoubtedly also become more important.

A young Indonesian representative who attended the first NNAF meeting in 1993 said the following: 'We cannot fight alone. We need your solidarity. That said, ultimately it is we ourselves who must stand at the forefront of the struggle. For if we are not willing to start this fight, nothing will happen.'

With the Stop Nuclear Exports Campaign as its center, the movement to prevent nuclear exports from Japan to Indonesia involved a large number of people. We who live in an export country felt that silence was not an option, knowing that activists and local citizens in Indonesia were risking their lives in the face of a military dictatorship allowed neither freedom of speech nor freedom of assembly. We must never forget how Japan has pushed Indonesia to import nuclear power heedless of the fact that the country has one of the highest occurrences of earthquakes and tsunamis in the world, a dreadful human rights record, and foreign debt exceeding 100 billion dollars. We must not allow this mistake to be repeated in the future.

Chapter 6

Taiwan: Aiming for the Complete Abandonment of Nuclear Power

Large-Scale Demonstrations

During the KMT (Chinese Nationalist Party) military dictatorship's 38 year period of martial law in Taiwan, under which citizens were prohibited from raising objections or criticism in any form, two nuclear reactors each were constructed at the sites of Jinshan, Guosheng , and Ma'anshan, totaling six reactors.

In 1987 martial law was repealed, and while the democratization struggle had been escalating for several years already, opposition to construction of the Fourth Nuclear Power Plant[12] proved a significant turning point for the movement. According to multiple organizations such as the Taiwan Environmental Protection Union, each passing year brought protest rallies on the scale of ten thousand people. The DPP, born from the democratization struggle, had

---

[12] Translator's Note: The Fourth Nuclear Power Plant is alternately referred to as Lungmen Nuclear Power Plant.

adopted an anti-nuclear platform. Yet the fate of the Fourth Nuclear Power Plant remained unclear. As the party in power shifted, parliamentary decisions to move forward with construction were frequently overturned, leading to uncertainty over whether this fourth site would be built. The problem of the Fourth Nuclear Power Plant became Taiwan's greatest political concern.

At the third NNAF meeting held in 1995 in Taiwan (including 32 participants from Japan), an incredible turnout of over 30,000 people participated in a large-scale demonstration, opposing construction of the Fourth Nuclear Power Plant and simultaneously demonstrating against France's nuclear weapons tests. During the demonstration, participants chanted 'End Nuclear Arms!' and 'Reject Nuclear Power!' And just before disbanding, they burned replicas of nuclear weapons and nuclear power plants in the middle of a main intersection. Overseas participants visited Lanyu (or Orchid) Island and Taiwan's first and second nuclear plants (at Jinshan and Guosheng), as well as an apartment building contaminated by radioactive materials. (This radiation originated from recycled building materials that had been contaminated by nuclear waste, that are confirmed to have exposed more than 15,000 people to radiation at 170 locations.) The forum then concluded with a protest within the perimeter of the proposed site of the Fourth Nuclear Power Plant, followed by a social gathering with Gongliao District residents and forum participants.

Lanyu Island, One Indigenous Community's Battle Against Nuclear Waste

During the third NNAF meeting in Taipei City, a Tau indigenous representative from Lanyu Island spoke in opposition to the

100,000 barrels of nuclear waste being brought onto the island for disposal, urging 'Don't make our island a nuclear waste disposal facility!' Hearing these words, the Tahitian participant rushed on stage to shake hands with the Lanyu representative. Tahitians not only wanted to be freed from nuclear testing, he said, but also wished to be freed from France's colonization.

The following year in 1996, Lanyu Island residents were able to stop additional import of nuclear waste onto their island by sheer physical resistance. In the early morning hours of April 26, the pier of the island was filled with 400 residents. Among these, some 70 protestors wore their Tao indigenous battle regalia and bore long spears. The transporting vessel was forced to retreat, and the barrels of nuclear waste were returned to Jinshan nuclear plant. Moreover, the islanders then negotiated with Taiwan Power Company to have the corporation remove an additional 100,000 barrels from the island that had been transported earlier, by the end of 2002. However, to this day, the corporation still has not fulfilled this promise.

NUCLEAR REACTOR EXPORTS FROM JAPAN

In Gongliao, the proposed site for the Fourth Nuclear Power Plant, a local referendum implemented in 1994 resulted in 96% opposition to power plant construction. Additionally, Taiwan's capital, Taipei, included a municipal referendum on the construction of the Fourth Nuclear Power Plant in its March 1996 presidential ballot. According to the referendum, some 52% of voters opposed construction.

Regardless, in October 1996, the bidding contest for the Fourth Nuclear Power Plant site went forward, with General

Electric winning the bid for the plant as a whole, while Hitachi and Toshiba were tapped to manufacture the nuclear reactor. In response, many people began to travel between Japan and Taiwan to address the dangers of nuclear construction. For example, city council members from Kashiwazaki, a Japanese town which hosted the same reactor model, the Advanced Boiling Water Reactor (ABWR), visited Taiwan and held press conferences to explain the design flaws and other problematic features of the ABWR model.

Local residents opposed to the nuclear construction deployed fishing boats and carried out marine training in preparation to block the port in order to prevent the nuclear components from being permitted entry. They carried out dramatic protests to express their staunch opposition to nuclear imports. For one such protest, a large dome-shaped model of a nuclear reactor with a Japanese flag painted on it was erected on a boat and set on fire in order to dramatize their opposition. In March 1997, some twenty representatives from Gongliao paid official visits to Ministry of International Trade and Industry officials, as well as to Toshiba and Hitachi executives, in order to protest and register their strong opposition to further nuclear expansion in Taiwan.

One concern regarding nuclear exports to Taiwan has been longstanding. Specifically, there is a question of whether Japan's nuclear exports are in violation of the Nuclear Non-Proliferation Treaty (NPT). Under the NPT, nations receiving nuclear materials are required to sign an additional protocol ensuring that neither facilities nor nuclear materials themselves will be utilized for nuclear weapons production. However, Japan and Taiwan do not have a formal diplomatic relationship and thus cannot sign an official bilateral agreement (as internationally required by the IAEA for export/import of all nuclear materials). In place of a bilateral agreement, Japan's Foreign Ministry employs simple verbal agreements that the US Department of State transmitted to the Japanese Embassy in US as the basis of its clearance under the NPT. This kind of unofficial verbal agreement, which does not even include the signature of the overseeing official, is simply a memo of correspondence and does not hold weight as a legal document. As such, it cannot provide an official guarantee that the NPT will be enforced, nor can it ensure there will be no transfer of nuclear materials for weapons production.

The verbal agreements between the two nations serve as a guarantee. However, the Japanese government appears dishonest in its assertion that a single-page document guarantees its adherence to NPT standards. Japanese campaigners organized several protests against the first Japanese export of nuclear power plants, including signature campaigns, product boycotts, assemblies, official queries in the legislature, participation in annual shareholders' meetings, and government negotiations, but none had the influence necessary to halt construction.

Despite a victory in the local residents' referendum and the clear opposition of the local governments and mayors (a situation that would be inconceivable in any other country), construction plans continued unabated. Although construction began in 1999, in 2000, the 50 year rule of the Chinese Nationalist Party's dictatorship came to an end, and Chen Shui-Bian, whose administration had made a public commitment to the 'abandonment of the Fourth Nuclear Power Plant', came into power. The Fourth Nuclear Power Plant Construction Review Committee debates continued for three months, and were even televised weekly in real-time broad-casts. The committee was comprised of equal numbers of proponents and opponents of nuclear power, and based on the committee's conclusion, the prime minister announced plans to terminate construction of the plant in October. However, a few months later, under the pressure of attacks from the opposition party, which stood to gain tremendous financial benefit from the construction of the nuclear plant, the political climate of Taiwan was thrust into confusion and turmoil. Towards the end of January 2001, Congress (which was dominated by the Chinese Nationalist Party) adopted a resolution to continue the construction of the plant. Eventually, Chen Shui-Bian conceded and construction resumed on February 14. The Japanese government, as if in anticipation of this result, promptly issued an export license on February 27.

During the 10th NNAF meeting, held in Taiwan in 2002, news broke that Tokyo Electric Power Company (TEPCO) had intentionally concealed an accident. During that gathering, twenty overseas representatives met with Taiwan's prime minister to discuss the hazards of nuclear power. Nevertheless, in 2003, Hitachi exported the first nuclear reactor to Taiwan from the port of Kure, and then in 2004, Toshiba exported the second nuclear reactor from Yokohama Port. The proposed site for the Fourth Nuclear Power Plant is adjacent to a beautiful coastal park, which was also where the Japanese Imperial Army first landed in Taiwan, an event commemorated at the site by an anti-Japanese monument. History repeated itself as the imported nuclear reactor from Japan arrived at the same location where Japan had initiated its colonial rule over Taiwan. Realizing Japan's true motives, the Taiwanese people began to refer to the nuclear exports as Japan's 'Second Invasion' of

Taiwan, and considered them a remnant of Japan's colonial invasion.

In 2003, the Gongliao Anti-Nuclear Self-Help Association issued the following statement: 'We strongly oppose Japan's act of "exporting pollution".… The export of nuclear plants from Japan is the equivalent of exporting hatred and terror into our hearts. We can also say this is the same as exporting tragedy.… On the other hand, the reason we were able to walk this tough path of nuclear opposition is because of the support and encouragement from our friends in Japan.'

Movie: *How Are You, Gongliao*

In 2005, Japan received a message from Taiwan. The message consisted of *How Are You, Gongliao*, a documentary film by a novice documentary director Tsui Shu-hsin, who moved to Gongliao and lived with the villagers for six years in order to make the film. This film, which was produced by the leading figure in Taiwan's documentary industry Wu Yi-Feng, was awarded Best Documentary Film in the 27th Golden Harvest Awards Film Festival.

Gongliao's people have been pushed around by upheavals in Taiwan's politics. Director Tsui Shu-hsin keenly observed the people of Gongliao. In the film, she portrays the people with their thoughts, anger, sorrow, and desires. She explained, 'During the filming, I encountered many people who [have now] passed away. Even now when looking back on these scenes, I cannot withhold my sorrow. For these people, rejecting nuclear power is their way of truly loving the land, loving the ocean, and loving their family. By portraying these people, I want their voices to reach out to everyone. Since Japan is the nation exporting these nuclear power plants, I would be very happy if the Japanese people could express interest in the plight of Gongliao's people.'

The chairman of the Gongliao Anti-Nuclear Self-Help Association, Wu Wen-Tong, together with Tsui Shu-hsin scheduled film screenings and dialogues with local communities across Japan between 2005 and 2006, including Osaka, Tokyo, Niigata, Kashiwazaki, Kitakyushu, Shimonoseki, Iwaishima, and Hiroshima. Perhaps surprisingly Iwaishima had the most dramatic reaction to the film. A 100 people had gathered in front of the public auditorium and the majority of them were women. They registered their empathy in agreeing with the film, responding enthusiastically, 'Yeah, yeah' and 'It's the same here!' to the protest scenes involving the dispatch of 100 fishing boats, the demonstrations against the electric company, and protests against the government officials. The people of Iwaishima had endured the same battles during the same 24 year period. Tsui Shu-hsin reported, 'The reaction here was more heartfelt than any of the 70 screenings in Taiwan. The people of Iwaishima's hearts came together with the people of Gongliao.'

The movie begins with a letter sent to Yuan, a young man who was falsely imprisoned, and ended in a climax with him finally receiving a prison furlough after 11 years of imprisonment. Instead of visiting his family, Yuan went straight to Fulong Station and met with Wu Wen-Tong as well as many other residents; embracing them tightly upon meeting.

Just before the start of the film screening in Iwaishima, Wu Wen-Tong received a call from Taiwan. 'In three days, Yuan will finally be released after being imprisoned for fourteen years and five months!' When Wu Wen-Tong announced this news, the crowd responded with thunderous applause.

Wu Wen-Tong remarked that it was as if he were in Gongliao, 'Iwaishima is just like Gongliao. I have been so encouraged by meeting everyone here in Iwaishima. Thanks to the film, I have been able to visit Iwaishima. I am very happy to be here today, and will continue to fight until the nuclear reactors are stopped. We must work together to stop both Kaminoseki, known as the last nuclear plant in Japan, and the Fourth Nuclear Power Plant, Taiwan's last nuclear plant together'.

On Communicating the Dangers of Earthquakes and Nuclear Power

From Japan, we have continued to relay information about the dangers earthquakes pose to nuclear power facilities. Many people from Taiwan attended the 12th NNAF meeting hosted at Kashiwazaki Kariwa in 2008. In 2010, Japanese scholars conducted a geological investigation on the site proposed for the Fourth Nuclear Power Plant and consequently discovered a new fault line on the location. At the 13th NNAF meeting hosted by Taiwan held shortly after this discovery, participants demanded that that the government standards for seismic retrofitting be improved. The construction of the Fourth Nuclear Power Plant was substantially delayed. In March and May 2010, there were accidental fires in the central control room resulting in a 28 hour long station blackout (full scale electricity loss). Furthermore, in 2011, it was revealed that the Fourth Nuclear Power Plant had 800 unapproved design changes.

I Am Human, I Am Anti-Nuclear

On March 20, 2011, 5000 people protested at Taipei City, demanding the closure of all operating nuclear reactors, as well as the immediate termination of construction of the Fourth Nuclear Power Plant. On April 30, some 15,000 people protested at many other locations throughout Taiwan. At the Taipei rally, two women from Fukushima took part and cautioned their Taiwanese counterparts, 'Do not repeat Fukushima'.

After the protests, many prominent individuals began to express their objections toward nuclear power, starting with well-known movie directors, actors, musicians, entertainers and cultural figures, as well as whistleblowers. In 2012, President Ma Ying-Jeou uttered the following political gaffe: 'I have never seen a person expressing their opposition to nuclear power'. In response, the 'I am human, I am anti-nuclear' campaign, initiated by a well-known film director, was launched.

On Ketagalan Boulevard in front of the Presidential Office Building, some 60 people staged a flash-mob style protest, yelling 'I am human, I am anti-nuclear', falling down on the street in the shape of the mandarin Chinese character for 'human' (人), and then quickly dispersing. For its second layer, the flash mob took aim at Taipei Main Station. Forming from pedestrians scattered across the bustling station, a mass of people suddenly joined together, forming the 人 character and yelling, 'I am human, I am anti-nuclear', and once again quickly dispersing. From this point on, individuals continued this protest by dropping to the ground and forming the 人 character. Others wrote 'Anti-Nuclear' on their fingers, forming them in the shape of the 人 character and taking photos and posting them to Facebook, triggering a chain reaction across social media sites. Artists also began to create posters to express their resistance to nuclear power.

On March 9, 2013, over 200,000 people staged a demonstration across different locations in Taiwan, known as the Taiwan Nuclear Termination Rally, with the stated goal of terminating all nuclear projects in Taiwan. Taiwan's total population is 23 million people, so if we were to apply the same ratio to Japan's population, participants in the demonstration would have been equivalent to one million people. Coordinated by more than 400 civil society groups, the organizers announced that the demonstration numbered some 120,000 people in Taipei City, 30,000 people in Taichung City, 70,000 people in Kaohsiung City, and several thousand people in Taitung City. The demonstrations continued well past dusk, occupying the streets through film screenings and musical performances. Many young people set up tents and stayed all night.

Non-Violent Direct Action

On March 8, 2014, another Nuclear Termination Protest Rally took place in the rain with around 130,000 participants. A car outfitted with megaphones guided the procession to the intersection in front of the Legislative Yuan (Taiwan's central government) in Taipei City. In a surprise tactic, the protest staff sealed off the area with yellow tape, and several thousand citizens stopped traffic by occupying the intersection. Meanwhile, the police were unable to react quickly enough to stop them. Protest staff wearing protective gear sounded a nuclear emergency siren and the protestors fell to the ground as if they were dead, demonstrating that in the event of an actual nuclear emergency, Taiwanese citizens would have no alternative but death.

This street occupation lasted 30 minutes, during which protestors shouted the slogan 'No Nuclear Exercise!' In mandarin

Chinese, this phrase involves a play on words: by substituting one character for another, a pun is created that cleverly incorporates both 'Non-Cooperation' and 'Anti-Nuclear'.

From March 18 to April 10, protestors had carried out an occupation of the Legislative Yuan in order to place pressure on its ratification of the Cross-Strait Service Trade Agreement with China. An occupation of this sort had previously been unheard of, and was dubbed the 'Sunflower Student Movement' by Taiwan's media, and eventually covered by media across the world. While it is true that the movement's main spokesperson was a student representative and the occupiers of the Legislative Yuan were predominantly young students, many civil society organizations were also supporting the movement. Every day, several thousand people surrounded the Legislative Yuan and formed a shield to prevent passage in or out of the building. On March 30, 500,000 people staged a rally there. Under Ma Ying-Jeou's administration, nuclear power had been promoted, along with other arbitrary decisions ignoring popular opinion.

In a season of relentless heat during the occupation of the Legislative Yuan, Lin Yi-Hsiung, who had become a symbol of the democratization struggle, announced his intention to sacrifice his life in order to abolish the Fourth Nuclear Power Plant, and went on indefinite hunger strike on April 22. Across the nation 126 anti-nuclear organizations, members of the National Nuclear Abolition Action Platform, responded to the hunger strike. On April 26, a mass demonstration of citizens conducted a sit-in in front of the Presidential Office Building and announced that the following day they would occupy the streets in support of the protest movement.

On the afternoon of April 27, 50,000 people started a protest march from the Presidential Office Building on Ketagalan

Boulevard, shouting the slogan 'Terminate Nuclear Power, Return Sovereignty to Citizens!' The marchers occupied Zhongxiao West Road, directly opposite Taipei Main Station. With the sheer mass of people, they were able to break through the areas sealed by the police and occupied eight lanes of traffic for 15 hours. In response to this series of intense protests, Ma Ying-Jeou agreed to a compromise, announcing an immediate freeze in construction and operation of both nuclear units at the Fourth Nuclear Power Plant, even though they were nearly completed. The media reported that 'The Fourth Nuclear Power Plant's operation would be halted under Ma Ying-Jeou's administration, and it will be left to the new administration in 2016 to make further decisions'. Lin Yi-Hsiung described it in this way: 'I have seen how the citizens of Taiwan have come together in a universal awakening because of the continuous protests that happened in recent months. If this awakening can be effectively organized and be channeled into an appropriate process, we can elevate the people's sovereign consciousness and raise the democratic capacity of the masses, and it could become an unstoppable force. With this strength, the ways of our current leaders will not be tolerated, and it will prevent any future dictatorship from ignoring the people's voice or from materializing in the future'.

In September 2014, sixty overseas attendees from the 16th NNAF meeting travelled by bus to visit the Taiwan Democratic Movement Museum located in Yilan County. Here, they had a meaningful exchange with Lin Yi-Hsiung and his wife, Fang Su-min. As one of the democratic leaders under the military dictatorship of the Chinese Nationalist Party during the 1970s and 80s, Yi-Hsiung was personally targeted by political terrorism. His mother and twin daughters were killed on February 28,

1980. Overcoming this painful experience, together with Fang he founded the Taiwan Social Movement Archives Center and the Taiwan Democratic Movement Museum.

What We Have Learned from Taiwan

The Taiwanese people have gathered in the tens of thousands for anti-nuclear protests and to occupy major streets in order to show their determination and assert their rights. In the Sunflower Student Movement, young people succeeded in historically unprecedented acts of resistance by occupying the Legislative Yuan. When we invited two young women participants in the Sunflower movement to Kansai, they shared valuable information with us. One of the women was in charge of public announcements during the Legislative Yuan occupation, while the other was in charge of the safety of young people who were occupying the Legislative Yuan.

When we asked how the young people in Taiwan were able to come together so energetically to coordinate such actions, they replied, 'We have seen the previous generation sacrificing themselves and fighting for their rights. Witnessing this, we felt we should do the same'. Their reasoning immediately called Lin Yi-Hsiung to mind. Lin had vowed to risk his life for an indefinite hunger strike, despite his advanced age. Of course, we are not strictly talking about famous protest figures even ordinary people have been able to exercise their political power as individual citizens.

During the occupation of the Legislative Yuan, the protest staff coordinated citizens to surround the perimeter of the building with tents to ensure that no one was harmed during the demonstration. Some individuals set up a soup kitchen and medical staff were deployed to care for the injured. Some individuals issued handmade newspapers to deliver important information, while others picked up trash to help maintain the area. Wheelchair-bound elderly participants even joined the protests, pulling their IV drips along with them. Artists showed their version of support for the young people by decorating the streets with their art. On holidays, these artists hosted handicraft workshops for children, decorating the streets with charming works of art that incorporated sunflowers.

Utilizing non-violent protest and with each individual offering their own skillset, the people of Taiwan have refused to give up. People who had fought through historical hardships and trauma under 38 years of martial law passed on memories and experiences, seeding their cause to the next generation, and also expressing the unique humor and warmth that was also present during that period. These lessons resonated with us through the inspiring conversation we shared with the two young activists.

When considering Japan's nuclear exports, it is impossible to skirt around the problems surrounding the Fourth Nuclear Power Plant, the first such export. Until this project is ultimately and completely terminated, we must join hands with the Taiwanese people and stand by them, learning together in order to move forward. We believe that the vision of a nuclear-free future will surely be inherited by generations to come.

Chapter 7

The Philippines: The Monster of Bataan and the Tenacious People Who Stopped It

THE NUCLEAR POWER PLANT THAT WAS NEVER SWITCHED ON

Even though the Philippines has a nuclear power plant, to this day it has never been turned on. This is the notorious Bataan Nuclear Power Plant, built by the American company Westinghouse on a beautiful cape in Morong, in Bataan province. Built under the military dictatorship of Ferdinand Marcos, it is an expression of the corruption, collusion, and social injustice that characterized his 20 year rule.

While the original budget for the plant was 500 million dollars, that sum had leapt to an excess of 2.2 billion by the time of its completion. Construction was begun in 1976 without once taking public opinion into account. After the Three Mile Island accident in 1979, work was temporarily suspended for safety inspections, turning up an astounding 4000 plus defects. As a member of the Philippine Congress said, 'Marcos and his supporters will go down in history as the men who built the most expensive and dangerous nuclear plant in the world, while saddling the Philippine people with so much debt that their heads will spin for generations.'

In 1981, with the Bataan facility as their common enemy, national networks, labor unions, and various other parties came together to form the Nuclear Free Philippines Coalition. Between 1982 and 1984, the people of Bataan engaged in grass-roots organizing, leading to the creation of the Nuclear Free Bataan Movement. Opposition to the plant grew nationwide. On International Human Rights Day (December 10) in 1984, the Nuclear Free Bataan Movement held a large and militant, province-wide people's strike, flooding the streets with thousands of people. Though Marcos's power had already grown weak by this time, he attempted to suppress the people's momentum with military force. The masses returned in 1985 with an even larger, three-day strike. 50,000 Bataan citizens packed the streets and faced down the military's tanks and armored vehicles. Construction of the Bataan plant was completed that year. The fuel rods were delivered, and were scheduled to be loaded before the end of the year. Due to the intensity of the opposition, however, they never were.

The following year, in 1986, fraud during the presidential election led to the eruption of long-standing anger at the Marcos regime and the emergence of the People Power Revolution. For those who had stood up against Marcos, the Bataan plant was a monstrous symbol of the evil and corruption of his military dictatorship. In April 1986, not long after the Marcos regime collapsed and Corazon Aquino became president, the Chernobyl disaster happened. The Aquino administration shut down the Bataan nuclear plant, and for 30 years since it has not once been operational.

Who Pays and Who Profits?

The Bataan Nuclear Power Plant was built by Westinghouse. The company reportedly paid Marcos and his supporters bribes on the order of 17 million dollars. Although the nuclear plant has never produced electricity, Westinghouse continues to demand payment. No Philippine citizen asked for the plant to be built, yet it is they who are saddled with this legacy of a dictator and his lackeys who are drunk with greed and in cahoots with multinational corporations.

In 1988, the Philippine government and Westinghouse began reconciliation talks. Following the talks and after numerous lawsuits, a deal was finally struck in 1995. The deal, highly unfavorable to the Philippine people, required the Philippines to pay a dizzying 300,000 USD per day. It wasn't until 2007—30 years later—that the country finally paid off this astounding sum and freed itself from the liability of the Bataan power plant.

In 1987, the Nuclear Free Philippines Coalition, which had been fighting against the transport of nuclear weapons into U.S. military bases in the Philippines, upped its game by becoming the driving force behind the addition of a 'no nuclear weapons' provision to the new national constitution. This provision became the cornerstone for the Philippine Senate's veto of the U.S. military basing agreement, which led to the shuttering of all American bases and the removal of all American troops from the Philippines by 1992.

ANTI-NUCLEAR ORGANIZING UNDER MILITARY DICTATORSHIP

After he became president, Fidel Valdez Ramos proposed that the Bataan plant be converted into a coal-fired power station. Then, in 1995, the administration changed its position. Now nuclear power was 'essential infrastructure for the country's development'. Included in its detailed nuclear energy policy were plans for a new nuclear power plant. The Secretary of Energy claimed, 'Since the Bataan plant was stopped in 1996, never once has the government really given up on nuclear energy'. In 1996, the 'Comprehensive Nuclear Power Program for the Philippines 2000' revealed that 10 sites around the Philippines were being considered for commercial reactors that were to be operational by 2020.

In response to this state of affairs, the fifth NNAF meeting, was held in the Philippines in 1997. Following an international conference in Manila, a public rally and a torch-lit procession were conducted in Bataan. At the rally, activists vividly described what it had been like fighting body and soul during the Marcos regime. In 1976, in an attempt to break the opposition, police opened fire on local protestors. Ernesto Nazareno, a Bataan local and one of the movement's leaders, was killed. Locals refused to cower after his death, however. Their commitment to the struggle only redoubled. Nevertheless, Marcos continued sending security troops to Bataan to force ahead the plant's construction. Large numbers of police and military were deployed to Bataan in 1981. When construction was completed in 1982, an entire military brigade was dispatched to the region. Despite severe military repression, locals continued to hold small discussions at the barangay (the smallest administrative division) level and to organize demonstrations. 200 students from across the Philippines converged on Bataan. For two months, they spoke with local residents to educate them about the dangers of nuclear energy. Due to the police and military presence, data and documents related to these meetings had to be buried and hidden in the ground.

THE BATAAN PLANT MUST NOT BE REVIVED

In 2009, influential pro-nuclear politicians proposed a congressional bill to revive the Bataan plant, leading to deliberations in the House of Representatives. Environmental groups, churches, scientists, farmers, fishermen, women, young people, and celebrities immediately came together to form the Network Opposed to the Bataan Nuclear Power Plant Revival and initiated protests and rallies around the Philippines. Bataan too saw many protests supported by the provincial governor and local government officials. People who had been active in the anti-nuclear movement at the end of the Marcos era dusted off their sandals and joined in. 'If the Bataan plant is making a comeback, then so are we,' they said energetically. 'In the name of our children, we will stop it from operating once again!' 'Southeast Asia has no need for nuclear power', intoned attendees from Malaysia and Indonesia at a rally in Quezon City, affirming support for their Filipino brothers and sisters.

In May of that year, three Bataan activists were falsely arrested by a group of 20 military and police officers armed with rifles and handguns. They were tortured while in custody. In June, at the Forum on ASEAN-Japan Cultural Relations in Tokyo, Philippine President Gloria Arroyo added fuel to the fire when she declared, 'We hope for the cooperation of the Japanese business community as our country pursues nuclear energy.' In the end, however, the Philippine people succeeded in crushing the Bataan plant's revival.

COLLABORATING WITH YOUNG ACTIVISTS

Citing electricity shortages, attempts to revive the Bataan facility are gaining force once again. Since the autumn of 2014, the government has carried out a series of public forums and site inspections to explore the possibility of reopening the plant. An assessment conducted by the government-owned National Power Corporation with the assistance of Korea Electric Power Corporation concluded that necessary repairs would cost more than one billion dollars. There has been a push to convert the Bataan facility into a coal-fired plant. But due to a high incidence of asthma and other harmful health effects from coal-fired plants elsewhere in the region, opposition to that proposal has been strong as well.

When we visited Bataan in January 2015, the plant was a far cry from the monstrous symbol of corruption and injustice that it had once been. The facility's grounds are serene and overgrown with greenery. Napping goats occupy the access roads. When the rare car approaches, they leisurely get up and relocate themselves to a quiet field. Not only can you go inside as far as the reactor core, which has never been loaded with fuel. You can even take photographs. They say that tourists come in droves from all over the country.

Young activists remain passionate about their work in Bataan even today. Older Philippine activists are constantly thinking about how to raise the next generation. They place deep trust in their juniors, and involve them in a wide variety of actions. Regarding the government's new pro-nuclear policy, one Philippine activist said, 'It's a provocation and insult to the Philippine people, who still have yet to recover from the damage of the Bataan plant.' Japan must not be allowed to support this 'provocation and insult' through nuclear exports. We have much to learn from the Philippine people.

Chapter 8

Thailand: Opposition to Nuclear Energy After the Fukushima Disaster

THE TENACIOUS PRO-NUCLEAR LOBBY

The Electricity Generating Authority of Thailand (EGAT) first officially announced plans to construct a nuclear power station in 1966. By this time, a smaller-scale 2000 MW research reactor had already been established near Bangkok. Initially, the anti-nuclear power movement in Thailand was part of the broader environmental movement, which included efforts to conserve forests and oppose dam construction projects. Although this initial attempt to establish the plant was suspended because of fluctuations in crude oil prices, in 1978 EGAT once again proposed nuclear power, and the National Assembly approved this plan. However, as a result of the Three Mile Island nuclear accident in 1979 and a global rise in anti-nuclear sentiments, Thai public opinion shifted against nuclear energy, seemingly putting an end to any further plans for nuclear plant construction.

Nevertheless, behind the scenes, EGAT continued to promote the idea of using nuclear power. They established the Office of Atomic Energy for Peace (OAEP) and in 1984, with financial support from the IAEA, conducted feasibility studies focusing on the economic benefits of nuclear energy. By 1992, EGAT and OAEP had drawn up blueprints and aimed to construct six reactors by 2001. In 1995, Thailand's premier research institution Chulalongkorn University established the Department of Nuclear Engineering in order to begin training personnel to staff Thailand's future nuclear program. That same year, the Ministry of Science and Technology began to actively push for the introduction of nuclear energy and revealed its plans to construct a nuclear plant in the south coast of the Gulf of Thailand.

In 1996, the Chief Minister of Science and Technology appointed a 'Committee of 21 Experts' to launch the nation's nuclear energy project. In January 1997, Canadian Prime Minister Jean Chretien visited Thailand to discuss sales of the CANDU Pressurized Heavy Water Reactor (PHWR). The government promoted its pro-nuclear agenda through the media and by sending key stakeholders to observe nuclear facilities in several European nations, and the nuclear program began to proceed at full tilt. After the government conducted feasibility studies at three sites in southern Thailand, the three sites previously designated to host nuclear reactors were made public. At this stage, EGAT began bombarding the public with pro-nuclear messages, investing significant funds in promotional television programs and commercials, as well as in comics, newspapers, and magazines. They also distributed promotional materials and videos to local schools within the communities designated to host nuclear reactors.

Initial Gathering at the Proposed Reactor Sites

When the sixth NNAF meeting was held in Thailand in 1998, representatives of Canada, the Netherlands, and several other Western countries attended as observers. NNAF participants also issued a joint statement strongly denouncing India and Pakistan's nuclear tests conducted earlier that year. After the main NNAF conference in Bangkok, the participants drove 600 kilometers to southern Thailand to visit two of the planned reactor sites. In the community of Chumphon, more than 200 local villagers gathered at a temple for Chumphon's first anti-nuclear rally. Residents voiced their concerns: 'Brokers also came to buy our land'; 'From about three years back, inspectors have come to this area, claiming they wanted to build facilities for children, like an amusement park'; 'They told me they would like to buy the deed of title for this land for only 2000 Bahts'. Local community members listened enthusiastically as the overseas guests discussed nuclear power's problems. Local residents of Surat Thani also held an assembly to meet with the NNAF guests.

Both communities historically have cleared out mangrove forests to make way for shrimp farms, which is oriented toward export to Japan. As we listened to the local activists speak, we keenly felt how unethical it is to push nuclear exports upon countries across Asia. As one local said, 'It is very hard to get people interested in or committed to the anti-nuclear movement in Thailand. This is because there are already so many urgent issues that need addressing, such as poverty, environmental destruction, and children's human rights.'

At the main NNAF gathering in Bangkok, residents of Ongkharak District in Nakhon Nayok Province near Bangkok, the community designated to site a research reactor, joined the conference. Many foreign companies had placed bids for the construction contract, and in 1996 the U.S. firm General Atomic was awarded the contract. Ongkharak District is a rich agricultural area that has a well-developed irrigation system and cultivates a significant amount of shrimp and freshwater fish. As four schools and a hospital lay within four kilometers of the site, community members were naturally concerned about the sudden decision to build a reactor here.

700 VILLAGERS FROM ONGKHARAK

Opposition to nuclear energy and the nuclear research reactor project was slow to develop amongst the Thai people, with the exception of local residents who were anxious about the projects and the risks they posed. The most recent radioactive accident in Thailand occurred in February 2000. News of the 'Cobalt-60 Radiation Exposure Accident' came as a tremendous shock, and was published on the front pages of newspapers across Thailand.

The accident occurred in a scrap metal factory in Samut Prakan, a region adjacent to Bangkok. Workers at the factory attempted to recycle several metal cylinders by cutting through them. Unbeknownst to the workers, these metal cylinders contained Cobalt-60 (manufactured by the German Siemens Corporation), exposure to which causes acute symptoms of radiation sickness. One after another, the workers collapsed from the radiation exposure and were sent to the hospital.

The accident resulted in three deaths and seven cases of acute radiation syndrome. Several 100 residents in the surrounding neighborhoods were also exposed. Three dogs also died in the area. Those who were exposed directly through handling the radioactive materials experienced immediate symptoms, including nausea, loss of appetite, hair loss, radioactive burns, bleeding from their gums, water blisters, enlargement of the lymph nodes, and a drastic reduction in white blood cells. Because the local hospital staff had not been trained in treating radiation exposure, the medical team was unfamiliar with how to respond. OAEP representatives rushed to the scene, but were criticized harshly by the media for their 'incompetence and inaction'. Despite the fact that the accident was caused by exposure to Gamma rays emitted from a medical radiation treatment machine manufactured by Siemens, local residents reported that a Japanese Toshiba machine had been placed directly next to the Siemens machine, and that if workers had cut into this for scrap metal, the Toshiba machine could have just as easily triggered the accident.

The following March, 700 villagers from Ongkharak travelled to Bangkok. They parked 10 buses in front of the Ministry of Science and Technology and staged fierce protests, stating that, 'Because the OAEP failed in its handling of the Cobalt-60 accident, it is impossible for them to safely operate a 10 MW research reactor.' Village representatives held private negotiations with the Deputy Minister of Science and Technology, compelling the minister to sign a memorandum stating that, 'As long as local residents oppose the project, the government will not permit construction of the research reactor.'

The First Forum For Nuclear Cooperation In Asia

In November 2000, the first Asian Nuclear Power Cooperation Forum opened in Bangkok. Japan took the lead in organizing the forum, which was attended by senior administrative officers and ministerial level governors from nine countries. NNAF, Japan issued a joint statement protesting the forum. Drafted by Thai anti-nuclear organizations, the statement raised crucial issues regarding the responsibilities of both the Japanese and Thai governments, 'From the dropping of atomic bombs to the Tokaimura criticality accident, Japan has been continuously exposed to a range of nuclear disasters. Seeing Japan lead these type of meetings surpasses the limits of irony'. Also: 'The Thai government brought up the cobalt-60 accident to argue that this forum would lead to prevention of these kind of accidents. This is unforgivable, considering that the OAEP has still not properly compensated people who lost their lives in the accident.' This statement was read by survivors of the cobalt-60 accident at a press conference in front of the elegant Sukhothai Hotel, where the Nuclear Cooperation Forum was held.

In the summer of 2003, the Thai government's environmental committee rejected the environmental impact assessment report as incompetent, and thus plans for building the Ongkharak Research Reactor became even less likely to go forward. In 2006, General Atomics cancelled its contract and the project was finally suspended. However, in September 2006, a *coup d'état* led to the establishment of a new administration and the return of the pro-nuclear lobby to power. In 2007, under the National Energy Development Plan, the government set a goal to construct a total of four nuclear power plants by 2021.

Demonstrations at Every Proposed Site

At the time of the Fukushima accident in 2011, the Thai government had proposed nuclear plants at eight sites: Trat, Chumphon, Nakhon Sawan, Surat Thani, Ubon Ratchathani, Khon Kaen, Kalasin, and Prachuap Khiri Khan. In March 2011, protests were held at each of these sites. In Kalasin, more than 2000 people demonstrated against EGAT's nuclear plans. Residents and citizens' groups from the eight sites then formed the People's Network Against Nukes. On March 17, the deputy prime minister of Thailand stated, 'As we do not wish to endanger citizens, the government is abandoning its plans to introduce nuclear power.' However, the prime minister later declared, 'Whether we pursue nuclear power or not is a decision that will be made in due time.' According to a public opinion poll, more than 80% of Thai citizens opposed nuclear power at that time.

Nevertheless, the Thai government continued to pursue its nuclear ambitions. In April, the government announced that it would delay the operation of its first reactor until between 2020 and 2023. Ironically, the Thai government held up Japan as an example to promote its pro-nuclear stance: 'Despite experiencing the world's worst nuclear energy catastrophe and still suffering from its impacts, the Japanese people still support nuclear energy.'

Because nuclear risks transcend national borders, the Thai people also harbor deep reservations about the Vietnamese government's attempts to establish its own nuclear industry. Environmental groups in Thailand sent a letter to the Vietnamese government, calling upon them to 'keep the Indochina peninsula nuclear-free'. In the fall of 2011, when the Japanese parliament was debating the nuclear cooperation agreement with Vietnam, citizen's groups in Thailand sent a flood of letters to Japan asking the Japanese government to 'please remember that, rather than exporting nuclear reactors to Vietnam, the Japanese government must provide relief and support to people still suffering losses and damages from the Fukushima accident'. We agree completely. Such clear-minded protests should be a wake-up call for any Japanese person.

Chapter 9

South Korea: Stopping Nuclear Development through Referendums

While South Korea was under military dictatorship, the government constructed nuclear power plants at four sites: Kori, Yeonggwang, Wolsong, and Uljin. In 1987, however, an anti-nuclear movement rose in the wake of the June Democracy Movement. Strong opposition across the country saved other locales from new nuclear power plants and nuclear waste disposal facilities.

In South Korea, like in Japan, pronuclear parties refer to the production of electricity through fission as 'atomic power' while only in South Korea do antinuclear parties prefer the term 'nuclear power'. In this text, we use 'nuclear power' to refer to South Korean fission production of electricity.

At the second NNAF meeting in 1994, the 36 participants from Japan joined the other conference attendees on a bus tour of South Korea. In Yeonggwang, we demonstrated with local farmers and fishermen against the nuclear power plant there. In Gwangju, we visited the gravesites of the victims of the Gwangju Democratic Uprising and protested in front of the American Culture Center. We held an assembly at Koseong and Cheongha, where people had beaten proposals for the construction of nuclear waste disposal sites. We also held assemblies and protests in Kori and Ulchin, which both have nuclear power stations. The Forum also served as an opportunity for activist and citizen groups from around South Korea to strengthen their ties. The nationwide network, National Headquarters for the Nuclear Power Eradication Movement launched soon after. Even beyond the duration of the NNAF meetings itself, many Japanese and South Korean anti-nuclear activists and citizens visited each other.

PROTEST AGAINST THE YEONGGWANG EXPANSION

Interactions between Japanese and South Korean anti-nuclear power groups date back to 1988 and the South Korean Anti-Pollution Civilian Movement Council. The two countries joined forces to petition against the construction of a third and fourth reactor at the Yeonggwang Nuclear Power Plant site (today called the Hanbit Nuclear Power Plant). Osaka Laborers Say No to Nuclear Power Plants was one of a number of Japanese groups who helped collect signatures for the petition, which was subsequently delivered to South Korea.

In Yeonggwang and South Jeolla Province, people are mainly Catholics or Won Buddhists. Citizen movements have been strong there for many years. Protests were waged against the construction of Yeonggwang reactors nos. 3 and 4 from the late 80s to the early 90s, and against nos. 5 and 6 beginning in the mid 90s. Numerous people from Japan participated in these protests.

In January 1996, 200 mainly Catholic residents occupied the Yeonggwang County Office and conducted a sit-in. In March, farmers in tractors converged on the front gate of the Yeonggwang plant, though they were stopped by the police. NNAF Japan issued a statement against the government's repression of the protests and arrests of local citizens. In April, 1400 Won-Buddhists initiated a coalition protest action and demonstration. Protests against the Yeonggwang expansion stretched as far away as Gwangju, located in the center of South Jeolla province. Nonetheless, under pressure from the Seoul central government and Korea Electric Power Corporation, the provincial governor ultimately approved reactors nos. 5 and 6.

CANCELLATION OF THE GULUP ISLAND NUCLEAR WASTE DUMP

In the 1990s, the pro-nuclear lobby's attempt to push through the building of a nuclear waste disposal facility invited violent protests from local citizens across South Korea. Nine out of nine rounds saw the pro-nuclear side falter. Which means, nine out of nine rounds went to the people.

Many Japanese participated in the protests on Anmyeondo Island and Deokjeok Island, located near Gulup Island. When Japanese anti-nuclear scientist Takagi Jinzaburō delivered a lecture on Deokjeok Island in February 1995, approximately 300 people attended. Deokjeok has a population of only 450, yet 350 riot police were stationed there during the event. On a wagon tour of the island led by young activists, Takagi commented on how much Deokjeok resembled Iwaishima, an island near Hiroshima that was fighting proposals for a nuclear power plant.

Despite deaths, serious injuries, and arrests, the people of Deokjeok continued to fight tooth and nail. Finally, that October, following massive protests in Incheon, a mainland city 80 kilometers from Deokjeok Island with a population of 2.3 million, plans for the waste disposal site were cancelled.

Opposition Against the Shin Kori Nuclear Power Plants

Just across the river from the Kori Nuclear Power Plant is the site of the Shin (New) Kori Nuclear Power Plant. When plans for their construction emerged, rising in opposition were not only local citizens but also the citizens of the one million-strong city of Ulsan, located less than 30 kilometers away. More than 5000 people attended the citizen's rally held in March of 1999. That May, 60 members of municipal, district, and county government took their struggle to Seoul. Japanese activists involved in the fight against the proposed Maki Nuclear Power Plant (in Niigata in northern Japan) held lectures to share their experiences with Koreans.

What's special about the anti-nuclear movement in Ulsan is that it was led by regional legislators and actively supported by large numbers of unionists from corporations like the Hyundai Motor Company. The cover of a color pamphlet published by the Strategic Committee of the Ulsan Metropolitan City, District, and County Government stated: 'Opposition to nuclear power is the natural duty of public servants.' However, due to the weak autonomy of local government in South Korea, the central government was able to force through construction at the Shin Kori site.

Protests Against Nuclear Waste Disposal Sites

Roundly defeated by the residents of Deokjeok Island and elsewhere in their attempts to build a nuclear waste disposal site, the pro-nuclear lobby disguised 250 billion won as 'regional assistance funds' that had been used by the end of June 2001. Expecting that protests would come to a head in the summer of 2001, we decided to hold the ninth NNAF meeting in South Korea. Government development committees were active in Yeonggwang, Gangjin, and Jindo Island (South Jeolla province), as well as in Gochang (North Jeolla). But, due to the strength of the anti-nuclear movement, ultimately none of the local governments were able to attract development.

Following a conference and street actions in Seoul, NNAF headed to Yeonggwang for the Yeonggwang People's Cultural Festival Against Nuclear Waste Disposal Sites. More than 800 people attended a protest in front of the Wolseong Nuclear Power Plant, where expansion plans had been proposed. A march in Ulsan against the Shin Kori plant attracted 500 people.

Democracy in Buan

Round 10 took place in Buan, in North Jeolla. In July of 2003, county governor Kim Jonggyu suddenly announced that he was seeking the construction of a nuclear waste disposal facility in his constituency. Under the watch of 8000 riot police, protestors renamed the city's main thoroughfare 'No Nukes People's Plaza' and held a nightly candlelight vigil of 2000–3000 people for 200 consecutive days. More than 10 times did 10,000–20,000 people collect for assemblies and marches. In November, an 'International Nuclear Waste Forum' opened in Buan.

The following is from *Buan Manifesto/Declaration*

In Buan they fought

Farmers rushed from their rice fields and into the streets

Fishermen threw down their nets and protested on the sea

Merchants closed their stores and defended the No Nukes People's Plaza

Taxi drivers abandoned their vehicles to block the highways

Mothers shaved their heads

Students refused to attend school

Grandmothers took up candles

Doctors rose up in their white gowns

Teachers bowed in entreaty 108 times before the National Assembly

Priests, educators, pastors, and monks fasted for over a month

Buan was united as one, without regard for differences in age or gender, in skill or talent, whether they were haves or have-nots

We devoted all we had to the struggle

On February 14, 2004, the citizens of Buan County voted in a autonomous citizens' referendum on the nuclear waste disposal facility and won. Of 52,000 eligible voters, 72% took part in the referendum, with 92% voting against construction. The popular referendum in Buan was modeled on that held in the township of Maki in Japan, which had defeated plans for a nuclear power plant. Just ahead of the referendum, donations were collected in Japan and taken by Maki residents to South Korea. It was the rare Buan citizen who sat out the struggle. Everyone was a hero. The people of Buan showed us what true democracy looks like.

REFERENDUM FRAUD IN GYEONGJU

Meanwhile, the pro-nuclear lobby continued looking for an opportunity. For three years, they had been sending, on average per year, 1000 people from various regions across South Korea to Japan for tours of the Rokkasho Nuclear Fuel Reprocessing Plant in Aomori prefecture. Activists from Aomori repeatedly visited South Korea to speak about the dangers of nuclear waste disposal sites. On November 2, 2005, during a referendum in Gyeongju sponsored by pro-nuclear parties, an 'overwhelming majority' supported construction, leading to plans for the building of a nuclear waste disposal site next to the Wolseong power station. The referendum was rife with fraud. Public officials extracted ballots from absentee voters, plying them with money and alcohol. A large number of absentee ballot requests were falsified, leading to an unbelievable 38% of the total submitted ballots coming from absentee voters.

SINCE THE FUKUSHIMA DAIICHI DISASTER

In 2011, South Korea had 21 nuclear reactors in operation, and another seven under construction. Shaken by the events at Fukushima Daiichi, 50 groups, including citizen organizations and labor unions, began a variety of actions under a new banner, Join Action for a Shift in Nuclear Power Policy and Support Group for Victims of the Fukushima Disaster (the name was later changed to Joint Action for a Nuclear Free Society). On April 4th, 'Hoping for a Nuclear Free World', a large candlelight vigil, was held in Samcheok, a candidate site for new nuclear power plant construction. On April 23rd, a demonstration in

Busan at the Kori plant demanded the closure of the aging no. one reactor.

The South Korean government held fast to its pro-nuclear position. In December 2011, the government unilaterally nominated two new sites for power plants: Samcheok in Gangwon province, and Yeongdeok in North Gyeongsang province. They also announced a goal of selling 80 reactors to foreign buyers by 2030, beginning with a deal with the United Arab Emirates. Not only did the South Korean government not back away from nuclear exports after the Fukushima disaster. They saw it as a chance to increase their market share.

January 2012 witnessed the inauguration of *No Nukes Newspaper* in South Korea. That March, the Pacific Basin Nuclear Conference in Busan, the Nuclear Industry Summit in Seoul, and the Nuclear Security Summit in Seoul all met with fierce opposition. The 15th NNAF meeting was held in Seoul, Samcheok, Yeongdeok, and Busan.

Events in Samcheok were truly inspirational. The rally venue was filled with men and women, both young and old, wearing racing bibs saying 'Fight Nuclear Power to the Death'. An antinuclear mass was held at a Catholic church that sits on an elevation overlooking the protest site. After the service was over, the congregation made its way down the narrow winding stairs to the plaza where we had gathered. More priests than one could count followed the people down the stairs in their vestments. Yet another phalanx of people descended after that. Over a 1000 people filled the plaza, which resounded with passionate appeals and the enthusiastic singing and dancing of young people. All of us from other countries gained courage by discovering so many allies in our struggle.

Referendum Victory in Samcheok

The NNAF gathering was just one of a number of large assemblies held in Samcheok to rally opposition against the construction of a new nuclear power plant. Though it failed, a referendum to recall the pro-nuclear mayor was held. Samcheok was also the starting point of a cross-country, anti-nuclear pilgrimage march. These activities culminated in the July 2014 election of an anti-nuclear mayor who publicly promised to hold a public referendum. Overcoming the government's attempts to obstruct the referendum, the vote was held on October 9, with 85% expressing their opposition to the city's attempts to attract nuclear development.

Seeking the referendum in Yeongdeok

In Yeongdeok, a petition for development was submitted with only the support of 399 residents from within the proposed construction area. There has never once been a public hearing on the matter. Civil servants from Yeongdeok county were sent to public relations seminars at Korea Hydro & Nuclear Power's (KHNP) Human Resource Development Institute, located on the grounds of the Kori Nuclear Power Plant. Higher-ranking officials were sent on tours of nuclear power plants overseas. To date, Yeongdeok county has sent nearly 90% of its 450 civil servants to nuclear development seminars. Five groups of 90 people each have participated in these three-day-two-night excursions.

Inspired by the Samcheok vote, 11 citizen and social organizations, including farmer and fishermen associations, formed the All-Yeongdeok People's Anti-Nuclear Solidarity

Group on March 2, 2015. This group paid close attention to Lee Jin Seop's lawsuit against his neighbor, the Kori plant, which he claimed was responsible for his wife's thyroid cancer. In October 2014, the Busan District Court ordered the Kori plant to pay 15 million won in damages. The judgment was important in that it officially recognized higher rates of cancer in regions with nuclear plants. All-Yeongdeok Solidarity held a demonstration on March 14. 'We have no choice but to express our will through a popular referendum,' cried the people, and on June 8 the Yeongdeok Nuclear Power Referendum Committee was formed.

SOLIDARITY WITH REGIONAL MOVEMENTS

At the first NNAF meeting in 1993 in Japan, the late Wonshik Kim from South Korea pointed out, 'It's only 220 kilometers from Kori to the Genkai Nuclear Power Plant in Kyushu. If an accident occurs at either facility, a wind speed of 7 m/s will ensure that the other facility will be irradiated within 30 hours. Japan and South Korea are bound by fate when it comes to a nuclear disaster. We, the people of Asia, must band together so tightly that the nuclear lobby doesn't stand a chance against us.' And indeed, solidarity and exchange between the South Korean and Japanese anti-nuclear movements has been robust ever since.

When the eighth NNAF meeting took place in Japan in 2000, a young Korean addressed the following words of encouragement to the residents they had met from Ibaraki, Fukushima, and Niigata. 'What has made you so strong? We thought about this, then found the answer on the hand-towels you gave us as souvenirs: To Be Against Nuclear Power is to

Be Human. Because you believed this, you had the strength to persevere and thrive. We would like to walk with you, along this human path.' It fills us with hope to know that there are young people who feel like this. Their words are a testimony to how deeply the regional movements in Japan have struck into a well of truth.

In 2000, we had gathered for a meeting at the Naraha Community Center in Fukushima. After the 2011 meltdowns, this building fell inside the exclusion zone. South Korea has become a country hell-bent on pursuing overseas nuclear exports. The time has come for us to not only treasure the bonds that we have already built, but to nurture even stronger ones to ensure that our countries do not victimize others through more nuclear exports in the future.

Chapter 10

We the People of the Global Nuclear Chain

Uranium Mining: The Case of Jabiluka, Australia

Let's begin with the 17 year old opposition to the Jabiluka Uranium Mine in Australia. Here is a concrete example of how the other countries' antinuclear movements have gained strength from Japanese support, which in turn can influence the course of uranium extraction.

Uranium mining in Australia has destroyed the environment and inflicted great suffering upon local peoples. Most acutely impacted are the Aborigines, who are further distressed by the fact that uranium extracted from their homeland is exported around the world and contributes to global nuclear weapons and nuclear energy problems.

While the mining itself takes place inside the Kakadu National Park, in Australia's Northern Territory, Jabiluka is an international issue. Even many Japanese people have gotten involved in campaigns against uranium mining. Not only has Jabiluka's development been strongly supported by Kansai Electric Power Company (KEPCO), with the expectation that much of the uranium would be exported to Japan. Japanese people were also moved by the fact that the mines would destroy the sacred lands of the Aborigines and would be located in the middle of a World Heritage area. NNAF, Japan helped raise consciousness about the issue in Japan by producing a Japanese version of David Bradbury's *Jabiluka* (1997) and selling postcards. We also attempted negotiations with KEPCO and KEPCO's largest shareholder, Osaka City.

Recognized by UNESCO for both its cultural and natural heritage, the Kakadu National Park is one of the few mixed World Heritage sites. At the 1998 UNESCO World Heritage Convention in Kyoto, concerned citizens from Japan and Australia assembled to inform the world about Jabiluka and urge for Kakadu to be put on the 'List of World Heritage in Danger'. Yvonne Margarula of the Mirrar Gundjeihmi people, who are the traditional owners of the Jabiluka lands, visited Japan and appealed for the mining to be stopped and the land to be returned to her people. Led by Hosakawa Kōmei of Kyoto Seika University, the Japanese contingent worked closely with Yvonne. Their activities were widely covered in newspapers and on television. As a result, the World Heritage Committee recommended the temporary suspension of mining operations at Jabiluka.

We went with Yvonne to protest at KEPCO headquarters in Osaka, but she was refused a meeting and turned away at the door. When she took one step up KEPCO's front stairs to stand behind a pillar and shelter herself from the cold breeze (Yvonne is from the tropics), a guard strode up to her and moved to push her back. 'Please do not trespass on KEPCO property,' he said. In a flash, Yvonne turned the tables. 'Then tell KEPCO to get off my land!' to which the crowded erupted in cheers.

Though the path to victory would be a long and arduous one, Australia's indigenous peoples eventually succeeded in halting mining operations at Jabiluka. Half of the uranium extracted from Australia is exported to Japan and South Korea, where it is used as fuel for nuclear power plants. We must not forget that our reliance on nuclear energy is one of the causes of Aboriginal people's suffering, and that the same tragedy has befallen the indigenous peoples of the United States and Canada.

After the Fukushima Daiichi disaster, Yvonne sent a letter to UN Secretary General Ban Ki-moon. She wrote about how shocked the Mirarr people were that the accident had been caused by uranium from their homeland, and appealed for an international ban on uranium mining. Alas, following the 2014 India-Australia nuclear cooperation agreement, Australia is today attempting to position itself as a major supplier to India's oversize, foreign-made, nuclear power plants.

When Indian Prime Minister Narendra Modi visited Canada in April 2015, he succeeded in getting the export ban on Canadian uranium to India lifted. It is projected that, over the next five years, 3220 tons of uranium will be imported by India. The nuclear fuel fabricated from this uranium will be used in reactors fabricated in Japan. Running all around the world, Japan buys up as much uranium as it can get its hands on—a mineral, that is, which is only good for two things: Nuclear fuel and weapons. Contamination from the fuel produced from this uranium has spread widely. How can we shrug our shoulders?

RARE EARTHS IN MALAYSIA

The refining of tin and rare earths also produces radioactive waste. That hasn't stopped Japan from being obsessed with obtaining rare earths, as they are essential for modern industry. During the country's era of high economic growth, the domestic processing of rare earths was strictly regulated in Japan. So instead, its industries engaged in what is known in Japan as 'pollution exporting', in which dangerous processing facilities are forced upon other, laxer Asian countries. This pattern continues today.

In 1984, Mitsubishi Chemical and local Malaysian interests created Asian Rare Earths (ARE), a joint venture company located in Bukit Merah village, in the tin-mining state of Perak. In 1984, radioactive exposure from thorium, a refining byproduct of the the rare earth extraction from monazite was reported. Large numbers of locals, including children, begin suffering poor health. Angry at the blatant contamination of their environment, locals form the Perak Anti-Radiation Committee sue ARE, which is ordered to cease operations by the regional High Court in October 1985. In 1987, however, the company is issued a permit by Malaysia's Atomic Energy Licensing Board to reopen. Later that year, in support of the beginning of Bukit Merah villagers' hearing, thousands of people march to the regional High Court from many kilometers away. They occupy the court grounds and protest for many days. In 1992, the High Court issues an injunction against ARE to shut down. The decision is remembered as a victory for one of the most tenacious struggles in Malayasian history. The Bukit Merah incident remains one of the main reasons why Malaysia has yet to succeed in building any nuclear power plants.

During the course of the long court battle, many villagers died, beginning with many children from leukemia. People continue to suffer from health problems today. Many Japanese, including former employees of Mitsubishi Chemical, have shown their solidarity with the victims of Bukit Merah by protesting against Mitsubishi Chemical. Dr. T. Jayabalan, a public health expert who dedicated himself to medical treatment in Bukit Merah, said at the first NNAF meeting in 1993, 'The village of Bukit Merah was created as a result of Japan's invasion of Malaysia during World War II. Then Japan returned, and went and created a plant without bothering to consult the people who lived there. The plant forced on the villagers lead particles, radioactive gas, and radioactive waste.'

Thirty years have passed since then. Opposition to rare earths processing facilities in Malaysia continues to expand. In the city of Kuantan, on the east coast of the Malay Peninsula, there are serious concerns that thorium and other byproducts from a large rare earth plant built by Australia's Lynas Corporation will cause environmental contamination. As almost all high-technology equipment requires rare earths, Japan seeks a stable supply of them. For the construction of the Lynas facility, the country loaned approximately 20 billion yen through the government-controlled Japan Oil, Gas and Metals National Corporation (JOGMEC). Japan has also signed a contract for the purchase of 8500 tons of rare earth every year over the next ten years (30% of domestic demand). Because the cost of processing rare earths is too expensive in countries with strict environmental regulations, the work is relocated to other countries. As a society that quickly buys and throws away electronic devices, we must beware of repeating past mistakes.

BANGLADESH

The Pakistan Atomic Energy Commission was established in 1955. In 1961, the country began considering the introduction of nuclear power plants. In 1963, land for potential construction was secured at Rooppur, located on the shore of the Padma River, a distributary of the Ganges, some 160 kilometers northwest of Dhaka.

Then in 1971, just as Belgium was about to finalize a deal for the Rooppur plant's construction, the War of Independence broke out. East Pakistan became Bangladesh, and in 1973 formed its own Atomic Energy Commission. The necessary legal framework for introducing nuclear energy was diligently set up, beginning with nuclear cooperation agreements with France in 1980 and the United States in 1981. The National Committee on the Rooppur Nuclear Power Plant was established in 1995. The following year, nuclear energy featured in the first National Energy Policy. In January 2005, Bangladesh signed a nuclear cooperation agreement with China, initiating a honeymoon period of nuclear relations between the two countries until 2009, when Russian cooperation was officially announced. The Russian and Bangladeshi governments vowed to have 2000 MW nuclear power plants operating in Bangladesh by 2017.

Cries against nuclear power have also been heard in Bangladesh. Politics alone explain Rooppur's selection as a site fifty years ago. To this day, no environmental impact assessment or other siting studies have been carried out, but this has not stopped the government from trying to build 2000 MW plants on the site. While water for the cooling system is to come from the Padma River, 75% of the river's water is consumed by neighboring India. Some people fear that this leaves too little water to safely operate a reactor. Furthermore, while Bangladesh just says yes to everything Russia puts on the table, the model of reactor Russia is planning to build, the VVER 1000, is outdated. Safety concerns led to a ban on its construction domestically in 2008.

What's more, Bangladesh is an economically vulnerable country. It lacks nuclear experts and engineers, and does not have the necessary industrial or transportation infrastructure. High-quality stainless steel sheets and piping, valves, pumps, etc. will likely all have to be imported. The cost defies imagination. There is also no regulatory framework or institution to ensure the safety of nuclear plants. The government keeps saying that Russia has agreed to dispose of all radioactive wastes, but Russia has clearly stated that no such agreement has been made.

Two representatives from Bangladesh attended the Lessons from the Citizens of Fukushima: A World Conference, held in Fukushima city in March 2015. During a panel discussion, a Russian representative said to the Bangladeshi representative sitting next to him, 'I feel horrible that my country is attempting to export nuclear plants to yours'. In the future, we are bound to see more attempts by nuclear exporters to sell their wares to countries that should be focusing their resources instead on protecting their citizens' health, raising their levels of education, and improving basic infrastructure. Just because a given country isn't being targeted specifically by the Japanese nuclear industry, doesn't mean its nuclear policies shouldn't be our concern.

PAKISTAN

After the first six sessions were held in East and Southeast Asia, NNAF finally arrived in South Asia with its seventh incarnation, in India in 1999. It was a tense time, since Pakistan had just conducted its first nuclear weapons test in May 1998, in response to India's second test two weeks before. With two neighboring nuclear weapons states facing down one another, those wishing to avoid nuclear catastrophe redoubled their solidarity.

Nuclear issues expert Zia Mian was scheduled to attend the NNAF in India from Pakistan. But as his visa was denied, Mian's national report on the situation in Pakistan was read in absentia. Titled 'Defense for Death,' his report began with the following paragraph: 'India and Pakistan are ruled by people who have no sense of reality. This applies to the prime ministers, military leaders, scientists, and bureaucracy of both countries. While the common people struggle with food, education, and shelter under conditions of extreme poverty, the elite spends obscene sums on weapons of mass destruction, beating their chests with pride in their ability and will to inflict the worst kind of cruelty.'

In South Asia, it is impossible to treat nuclear energy and nuclear weapons as separate issues. Everyone knows they are connected. As of 2014, there were three nuclear power plants operating in Pakistan, with another two under construction and two more under consideration. The 137 MW Karachi Nuclear Power Complex, composed of CANDU reactors supplied by Canada, went online in 1972. In 1974, India conducted its first nuclear weapons test using materials sourced from its own CANDU reactors, which it had obtained from Canada for ostensibly 'peaceful purposes'. A shocked Canada froze its nuclear cooperation agreement with India. The Indian test eventually led to the creation of the Nuclear Suppliers Group (NSG), which aims to prevent arms proliferation by controlling the export of nuclear technology and fissile materials. Canada also proposed strengthened safeguards against Pakistan. But as the Pakistani government refused to accept them, Canada ultimately cancelled their cooperation agreement.

Pakistan subsequently succeeded in creating a uranium enrichment plant. As Pakistan had elected not to sign the NPT, it could not count on support from other countries for the development of large-scale reactors. It looked like Pakistan would have to develop them on its own—until it signed an agreement with China in 1986. The first Chinese-made reactor began operation in 2000 at the Chashma Nuclear Power Plant, in the province of Punjab, with a second to follow in 2011. Number 3 and 4 are currently under construction.

Ahead of the NSG plenary meeting in September 2008, the United States suddenly began courting India. At the meeting, the NSG approved to exempt India from the provision that bans nuclear imports to non-NPT states, leading Pakistan to ask the international community for the same. At the opening ceremony of the no. 2 Chashma reactor in May 2011, Pakistani Prime Minister Yousaf Raza Gillani delivered a speech saying, 'When it comes to the transfer of technologies related to the peaceful use of nuclear energy, the international community must refrain from discriminating against certain countries.' Japan's attempts to export nuclear power plants to India only serve to exacerbate arms tensions in South Asia.

Jordan and the United Arab Emirates

The nuclear cooperation agreement between Japan and Jordan went into effect in February 2012. That with the United Arab Emirates (UAE) was signed when Prime Minister Abe Shinzō visited the Middle East in May 2013, and went into effect in July 2014. As for Saudi Arabia, after a basic agreement was made during administrative level discussions during Abe's visit, Japan's Ministry of Foreign Affairs announced in December 2013 that official negotiations on the agreement had begun. Abe himself is personally advocating for nuclear sales to the Middle East. Nuclear energy is new to each of these countries.

Jordan is a small nation surrounded by Israel, Syria, Iraq, Egypt, and Saudi Arabia. It is highly vulnerable to acts of terrorism. Though in the Middle East, Jordan is not an oil exporter, so its economy is weak. It experiences many earthquakes. Limited public access to information makes democratic participation difficult. And the list of problems goes on. The proposed location for a nuclear plant, Al Amra, is located in the desert 85 km from the nation's capital of Amman.

In Japan in January 2012, during Global Conference for a Nuclear Power Free World, a member of Jordan's House of Representatives and a Jordanian lawyer talked about the various problems that countries in the Middle East face. Despite the fact that 80 out of 120 members of Jordan's House of Representatives are opposed to nuclear energy, the lower house has practically no power or influence over the government's decisions. They said that the lower house was not even informed about the signing of the nuclear cooperation agreement with Japan. The Jordanian participants also told us about how citizens are worried about the vulnerability of nuclear power plants to attack, considering Israel's airstrikes against the Osirak reactor in Iraq and a nuclear facility in Syria, as well as threats to the Dimona reactor in Israel.

In April 2012, the Jordan Atomic Energy Commission shortlisted two companies, Atmea (a joint venture between Mitsubishi Heavy Industries and France's Areva) and Russia's Rosatom, for consideration. That May, the Parliament of Jordan ordered the temporary suspension of all plans relating to nuclear power plants construction. Nonetheless, in September 2013, the contract was awarded to Rosatom. Reactor no. 1 is projected to be online in 2023.

In the UAE, two reactors are under construction, which will amount to 2800 MW. Two more are planned, for another 2800 MW. Under consideration is yet another plant, amounting to yet 14400 MW. The UAE government's 2008 plans thus identified 14 reactors totaling 20000 MW. For the construction of the first four reactors, a South Korean consortium was selected in December 2009. Areva, Hitachi, and GE's failure to secure the contract was a major shock to the global nuclear industry. South Korea's low pricing, military backing, human resource and training support, and willingness to accept all of the UAE's conditions, as well as its sales prowess, were said to be the reasons for its bid's success. South Korea's example is probably closely being followed by the Japanese industry. A groundbreaking ceremony for the Barakah Nuclear Power Plant was held on March 14th, 2011. Construction of unit no. 1 started in July 2012, and no. 2 in May 2013. Unit no. 1 should start supplying electricity in 2017.

The UAE is made up of seven emirates, each of which is ruled by an absolute, hereditary monarchy. Dubai is the most famous of these emirates. Citizens do not have the right to vote and political parties are banned. Political power is allotted to only a few thousand people directly selected by the Emir. Only 20% of UAE's residents are actual citizens. The rest are foreign laborers from other Arab countries and South Asia. Criticism of the Emir is forbidden.

Exporting nuclear technologies and materials to the Middle East is rife with problems. The questionable safety of nuclear power plants, the risk of earthquakes, the lack of transparency, and limitations on free speech make the projects in themselves problematic. But that's not all. Given Israel's refusal to admit to possessing nuclear weapons, other Middle Eastern countries' desire to pursue nuclear weapons and nuclear energy is only natural. This is surely a good thing for those who sell nuclear power plants, but we cannot stand by and let governments make all the decisions. What does enabling the proliferation of nuclear energy across the Middle East really mean? It's high time we give this question serious thought.

CHINA

As of December 2014, there were 20 nuclear reactors operating in China, with another 29 under construction, 59 planned, and 118 under consideration. The Fukushima Daiichi accident led to safety inspections and the review and temporary suspension of approval of all new construction plans. However, in October 2012, the suspensions were lifted and new construction recommenced. With its eye on nuclear exports, China is developing its own reactors. Nuclear power plants built by the Chinese are already in operation in Pakistan. International business deals with the United Kingdom, Romania, and Argentina are in the planning stages.

Considering the above, how could one not include China from any discussion about nuclear power in Asia? Yet, there are serious obstacles to linking up with China's antinuclear movement. Chinese policy advisor Wen Bo, who attended the ninth NNAF meeting in South Korea in 2001, noting the irony of Chinese media silence on domestic power plants, waste, and uranium mining, published this statement in 2008:

> In China, the media reported both the Chernobyl and Fukushima disasters; antinuclear protests in Taiwan and protests against nuclear waste transport in Germany. Yet, they remain silent on domestic nuclear issues. As such, most Chinese have little understanding of threats posed by domestic nuclear issues, or even that they have a right to speak out against them.

Wen Bo continued:

> Even as Chinese civil society and environmental groups have expanded massively these fifteen years, they have not focused on nuclear power plants or radioactive waste. Online campaigns are the mainstream of current anti-nuclear efforts in China.

Noting the prevalence of 'not in my backyard (NIMBY)' type arguments against industrial facilities, he suggested that such grassroots campaigns would help foster the birth of anti-nuclear power groups. Moreover, citing the radioactive contamination accident at the Gansu uranium mines, he added, 'Most people think that halting the explosive growth of nuclear power in China is just too difficult. Yet due to the spread of internet-based monitoring and people's growing awareness, problems will not be as easy to cover up as they were in the past.'

We Bo's words are from 2008. Given the eye-popping scale of Chinese nuclear expansion, it is hard to be so optimistic. But we can do our part to help locate the fissures by disseminating information in Chinese.

RADIOACTIVE WASTE KNOWS NO BORDERS: THE CFS INITIATIVE IN MONGOLIA

On May 2011, just two months after the Tōhoku earthquake and tsunami, the front page of the *Mainichi Shimbun* carried an unbelievable article. It began, 'Since last autumn, METI has collaborated with the U.S. Department of Energy, with secret plans to build the first international spent nuclear fuel storage and disposal facility in Mongolia. For Japan and the U.S., who lack domestic disposal sites, the plan is designed to counterbalance Russia and France's aggressive nuclear technology sales.' Both Russia and France, have sought to make their nuclear energy more attractive by bundling reactors and waste disposal plans together. After hearing this shocking news, the Mongolian people erupted in a chorus of protest.

'Mongolia will not accept nuclear waste from foreign countries,' said the country's president, dampening the flames. Also implied in this, however, was, 'We will accept radioactive wastes from inside Mongolia'. Thus, the Comprehensive Nuclear Fuel Service Initiative, or CFS, was announced. The initiative allows for spent nuclear fuel from foreign nuclear power plants to be accepted as Mongolian waste if that fuel was produced from uranium mined in Mongolia. In return, nuclear technology and know-how will be shared with Mongolia, which is reputed to have plans for introducing nuclear power.

The largest uranium deposits in the world are in Mongolia. Attempts to develop mines through projects backed financially by Areva and Mitsubishi have already caused environmental contamination. Mongolia's nomads have fiercely protested and even occupied the mining sites.

In the past, when plans to export radioactive waste from Taiwan to North Korea came to light, there were protests in Taiwan by South Korean environmentalists. To our knowledge, the final resting place for decontaminated materials from Fukushima is yet to be determined. How shameful, in the eyes of the world, that Japan is pushing for the CFS initiative as a response to the Fukushima disaster.

CONCLUSION: THE GROWTH OF NUCLEAR POWER

The above cases represent just some of the many countries attempting to adopt nuclear energy. The further proliferation of nuclear power plants across the world is a growing possibility. Considering that the countries that are now attempting to introduce nuclear plants are the countries that originally had no reason to even consider them, it seems clear that nuclear energy is a commodity that can only be sold to such countries.

The time has come for the citizens of nuclear export countries to band together. On the one hand, there is the general problem, experienced many times over, of multinational corporations thrusting their way into and wreaking havoc in foreign countries. On the other, the specific problem of nuclear energy exports and nuclear weapons. We must keep in mind both perspectives as we continue to grapple with the problem of nuclear power.

Afterword

Today two of every three nuclear power plants are built in Asia. Asian governments are joining forces with the governments and industries of traditional nuclear energy states. Nuclear power is now being promoted in countries with less freedom of expression, less political stability, and less reliable infrastructure.

People have long claimed that nuclear energy is incompatible with democracy. That has never been truer than it is today. In the 1990s, it was often said that the nuclear industry was in a state of decline. Companies have found it difficult to build new power plants in their home countries. The industry's heyday was over, and its pulse grew faint. The only way to guarantee the viability of nuclear technologies was to export them. It's hard to believe that an industry dependent wholly on foreign markets has much of a future. Nonetheless, that is how the nuclear industry has managed to survive. Even harder to believe is that these developments would intensify after the Fukushima Daiichi disaster, but that is exactly what has happened.

During the first NNAF meeting in 1993, a South Korean participant said as follows:

> In my country there is a saying that goes "Simply starting is getting halfway to the goal." But there is also another saying, "On a road of 1000 leagues, 999 might as well only be halfway." The road ahead of us is long and hard. It is filled with every kind of obstacle. Until we overcome these, let our hearts be as one. To hasten the day of our victory, let us fight as one.

NNAF has provided us repeated opportunities to exchange information, share experiences, and coordinate actions. Through trial and error, over time we have created a network that is both tight and equitable. Led by residents of places with nuclear power plants, or with plans for nuclear power plants, or from places threatened by potential waste disposal facilities, a large number of people have attended NNAF through the years. We learn from one another and encourage one another. We keep in touch beyond the annual conference, sharing information and visiting one another's countries. A nuclear industry paper was kind enough to credit us with 'exporting antinuclear movements across Asia'.

As Corazon Fabros of the Philippines has said, 'NNAF is not just for exchange between urban activists. It is a network that cherishes local peoples as well. Local peoples, in turn, gain strength when the forum is held. That is why NNAF has continued for 22 years. And that is why it must continue into the future. Indonesia, Thailand, and the Philippines have all succeeded in stopping the development of nuclear power. In each case, NNAF's contribution has been significant.'

This book was written as an introduction to the history of anti-nuclear power movements in Asia and the layers of interaction between different groups. It was also written with the belief that such a text would contribute to further strengthening solidarity. Pitted battles that the mass media fails to report, declarations buried with the passing of time, episodes born spontaneously from face to face interactions—we wanted to make sure that such things are not forgotten. We wanted to record them so that they may serve as inspiration for alliances to come.

The path has been long, but we have come a long way. Our goal is to put an end to nuclear exports by attacking the industry from both sides: From the countries where the exporters are based and from the countries where the importers reside. The fight against nuclear power in Asia is also a fight for democracy. An Asia without nuclear weapons or power plants is an Asia for the people. In the name of this non-nuclear Asia, let us continue to work to grow and strengthen the bonds between us.

Since we began our activities in 1993, many important friends in the fight for a non-nuclear Asia have passed away. Kim Wonsik, a longtime organizer of international solidarity between local grassroots movements, was one of the first proponents of NNAF and a man we could depend on like a father. Miyajima Nobuo was a wonderful teacher and a passionate comrade in arms. Ōba Satomi, director of Plutonium Action Hiroshima, taught us that true friendship with the people of Asia was a real possibility. He Zhao Ming of Taiwan, with love for his motherland and devotion to the anti-nuclear movement, gave his all to bringing together the movements in Taiwan and Japan. One after another, our friends from Gongliao, site of Taiwan's Fourth Nuclear Power Plant, have passed on. Dan Ginlin, beloved by the people of Iwaishima in Japan and who was always eager to involve us with young activists in Taiwan, passed away in 2013 at the young age of 30. Were it not for courageous individuals like these, the nuclear arms and energy situation in Asia would be far more difficult than it already is.

On February 2nd, 2005, the Global Marketing and Sales Division of the Japan Atomic Industrial Forum released a report titled 'Recent Trends in Global Nuclear Energy Development'. It claimed that 435 nuclear power reactors were operating, 72 were under construction, and 174 were in the planning stages, spanning 31 different countries and regions. Don't let these numbers make you lose spirit. Because behind every reactor is a community with a deep and expansive history made up of people who are ready to risk their lives to defend their homes.

Only by knowing these histories, honoring individual struggles, and understanding current conditions will the ways to continue the fight be visible. We may be able to meet one another in person only on rare occasions, so it is essential that we keep courage and spirit by remembering that our foreign brothers and sisters are continuing their fight back home. May the way forward be paved both with respect for the relationships we have already built and for the widening solidarity between grassroots groups across national borders.

An Update on the Developments in Nuclear Asia since 2015

After this book was published in Japan in 2015, a number of developments in people's movements and the nuclear industry have transpired in the Asian region.

Even as India and Turkey continued to reinforce and advance their reactor construction plans, democratization movements in Taiwan and South Korea have birthed new leadership and elected to abandon nuclear energy. In Vietnam, where the administration sought its first nuclear construction plans, the government has decided to scrap its nuclear development plans. To better assist readers in linking the situation in 2018 with the contents of the book, we provide a brief overview of these sweeping changes since the initial publication of the book.

INDIA

December 12, 2015: During Prime Minister Abe's annual bilateral summit with Prime Minister Modi in India, the two leaders agreed to 'Agreement in Principle' on the Indo-Japan Nuclear Cooperation Agreement (IJNA). On the same day, residents of Jaitapur which is slated for reactor construction under the IJNA, staged a massive protest. Using 'jail bharo', a non-violent direct action in which protestors attempt to jam the jails by spontaneously seeking to be arrested, upwards of 1000 people were arrested.

March 28, 2016: Together with NNAF, Japan, Indian civil society representatives lobbied Japan's Ministry of Foreign Affairs (MOFA) in Tokyo. MOFA bureaucrats responded that, 'We were unaware that there was any repression against Indian citizens in relation to nuclear reactor construction, nor that protestors had been killed. However, these are the internal affairs of the Indian government.'

October 2, 2016: (Memorial of Gandhi's Birthday) Major public gathering in Jaitapur.

November 2016: Joint international appeal to oppose the Indo-Japan Nuclear Cooperation Agreement, signed by 455 organizations in 31 nations.

November 11, 2016: Japan and India sign the IJNA.

February 24–June 7, 2017: 'Oppose the IJNA Campaign' promoted across the globe. IJNA debated and ultimately ratified by the Japanese Parliament on June 7.

May 18, 2017: India's National Green Tribunal ruled against granting environmental clearance to the Mithi Virdi nuclear reactor construction project and the project was shelved.

August 20, 2017: Major public gathering in Jaitapur, demonstration and 'Jail Bharo'.

March 10, 2018: Protests against French Prime Minister Emmanuel Macron's state visit to India, public gathering and demonstration in Jaitapur.

August 27, 2018: Major public gathering in Jaitapur, demonstration and 'Jail Bharo'.

TURKEY

April 16, 2017: National Referendum held on increasing presidential authority through a Constitutional amendment. Approved by 51% of the electorate.

February 6, 2018: Public hearing for the Sinop reactor project held; only pro-nuclear factions allowed to attend. Local citizens who attempted to enter the hearing were removed by the police.

April 3, 2018: President Erdoğan and Prime Minister Putin attend groundbreaking ceremony for the Akkuyu Nuclear Plant .

April 22, 2018: Mass public gathering and demonstration held annually in Sinop cancelled by the Turkish authorities for the first time.

April 24, 2018: Japan's Itochu Corporation withdraws from the Sinop nuclear reactor project. As a result, the entire project is likely to be delayed. (Due to a significant increase in the cost of the safety measures, the total project cost has swelled to more than double the originally budgeted amount, and is now anticipated at 5 trillion yen for four reactors). Moreover, on December 4, 2018, Mitsubishi Heavy Industries, which had led the Sinop NPP project consortium, announced that it was abandoning the Sinop nuclear project due to soaring costs.

VIETNAM

November 22, 2016: The Vietnamese government announced its decision to suspend its nuclear reactor construction plans. Among the reasons provided for cancelling its nuclear plans, the government mentioned concerns about the profitability of the reactor project due to the increased costs for safety measures, financial difficulties, and lack of trained personnel and nuclear engineers required for adopting nuclear energy.

TAIWAN

January 11, 2017: The revised Electricity Utility Act (or the De-Nuclearization Act), proposed by Prime Minister Tsai Ing-wen's administration was approved by the Legislative Yuan. The revised version of the law includes the provision that 'All nuclear reactors must cease operation by 2025'.

July 4, 2018: Roughly 80 unused fuel-rod bundles (out of a total of 1744) designated for fueling the Fourth Nuclear Power Plant were removed from Taiwan and transported to the US (as the first shipment).

PHILIPPINES

November 2016: President Duterte approves procedures toward initiating the nation's nuclear power plants.

November 2017: Russian State Nuclear Corporation Rosatom signs a Memorandum of Understanding with the Philippine government toward cooperation in nuclear technology.

November 2018: The Philippines hosted the 25th anniversary gathering of NNAF.

SOUTH KOREA

November 11–12, 2015: Residents of Yeongdeok conducted an autonomous citizen's referendum on the need for construction of a new nuclear power plant and successfully opposed the construction.

September 12, 2016: Gyeongju Earthquake (With a magnitude of 5.8, this was the strongest earthquake in Korea's recorded history).

June 19, 2017: At the ceremony for the permanent suspension of Kori Unit 1, President Moon Jae-in announced his 'Nuclear Abandonment Declaration'. He also promoted the Nuclear Safety Commission to be appointed directly under the president's supervision, completely shelved all plans for new reactor construction, suspended all lifetime extensions for ageing reactors, announced the early closure of Wolseong Unit 1, and declared that Korea would establish a roadmap toward rapidly abandoning nuclear weapons.

June 15, 2018: Decommissioning of Wolseong Nuclear Power Plant Unit 1 announced. Further, official announcement that construction plans for four new reactors at Samcheok and Yeongdeok had been cancelled.

AUSTRALIA

November 6, 2016: In South Australia, a 350-member citizen's jury ruled against plans to construct a radioactive waste disposal site. (The site had been proposed to accept nuclear waste from Japan, South Korea, and Taiwan, among others).

MALAYSIA

September 18, 2018: Prime Minister Mahathir announces that 'We (Malaysia) will not choose nuclear power'.

BANGLADESH

November 30, 2017: Construction on the nation's first reactor, Rooppur Nuclear Plant, officially initiated. According to official reports, Bangladesh is importing two Russian-made improved type Pressurized Water Reactors (1200 MW each), with Unit 1 slated for start-up in 2023 and Unit 2 in 2025.

Having engaged with the issue of nuclear power in Asia since the 1980s, our confidence today is absolute. The tide has definitively turned. Inexpensive, clean, and safe forms of renewable energy have increased rapidly. They now account for some 25% of the world's total energy generation, and more than twice the energy produced by nuclear power plants. As anyone can plainly see, nuclear power is nothing but a dangerous and cumbersome relic from the past. The embers of those who support the nuclear industry may continue to smolder, but it is impossible to imagine nuclear power being allowed to captivate society ever again.